TERRAIN

TERRAIN

TRAVELS THROUGH
A DEEP LANDSCAPE

GEOFF CHAPPLE

RANDOM HOUSE
NEW ZEALAND

RANDOM HOUSE

UK | USA | Canada | Ireland | Australia
India | New Zealand | South Africa | China

Random House is an imprint of the Penguin Random House group of companies, whose addresses can be found at global.penguinrandomhouse.com.

First published by Penguin Random House New Zealand, 2015

10 9 8 7 6 5 4 3 2 1

Text copyright © Geoff Chapple 2015

The moral right of the author has been asserted.

All rights reserved. Without limiting the rights under copyright reserved above, no part of this publication may be reproduced, stored in or introduced into a retrieval system, or transmitted, in any form or by any means (electronic, mechanical, photocopying, recording or otherwise), without the prior written permission of both the copyright owner and the above publisher of this book.

Cover and text design by Shaun Jury © Penguin Random House New Zealand
Illustrations as credited pages 267–68
Landscape profile page 5 by Geographx
Printed and bound in Australia by Griffin Press, an Accredited ISO AS/NZS 14001 Environmental Management Systems Printer

A catalogue record for this book is available from the National Library of New Zealand.

ISBN 978 1 77553 679 6
eISBN 978 1 77553 680 2

penguinrandomhouse.co.nz

For Miriam

CONTENTS

Northland — the allochthon	9
Auckland — the volcanic field	47
East Coast — subduction	71
Taupo — calderas	113
Wellington — earthquakes and fossils	143
Marlborough Sounds and the Red Hills — ophiolites	171
Westland — the alpine fault	197
Southland — terranes	227
Acknowledgements	263
Image credits	267
About the author	269
Geological timescale	270
Location map	271

NORTHLAND
The allochthon

It took a while to get used to the geologists' small talk. In the course of a year, I'd hear a lot of it, but Mike Isaac was the first. Coming down to the Cape Reinga lighthouse, he scuffed the distinctly rutilant gravel path and mused

— Red chert. From the McCallum Brothers Quarry on Karamuramu Island in the Hauraki Gulf.

We walked on down to the flat apron surrounding the Cape Reinga lighthouse and he wiped a Teva-sandalled foot across the flagstones.

— More geological pollution.

It was a phrase geologists tend to use when some human agency has disguised the original outcrop with imported rock.

— Limestone, said Mike, from the Paradise Quarry near Whangarei. Pity it's not a better colour. A bit grey-green, a bit cold-looking. If it was a nice warm Cotswold colour this would be a very popular building stone.

Mike's inquisition of the surrounding rocks never stopped. It was an eerie skill, far more than knowing the source quarries and the uses a human economy had impressed upon the stone. The rocks yielded clues of crystal and grain size, of the fractional melt that gives the hardest rocks their character as they come up from the mantle. Mike went across to the rock wall that edged the apron, looked it over and pronounced gabbro, dolerite and basalt boulders, probably trucked in from Larmer Road Quarry at Kaitaia.

— You can see the crystals in the gabbro are much bigger. Same chemistry as the basalt, but it's cooled a lot slower. The crystals have time to separate out and form.

Such was the chatter at the Cape Reinga light, but I'd wanted something else from Mike. I wanted to know about what geologists in their more light-hearted moments call 'the Great Thon'. Slumbering beneath the various lapidary distractions as we came onto the Cape was the Great Thon. That's why we were here, and even a geological newbie like me could feel its power. The land ended in stunted vegetation and steep cliffs of basalt. This was where, according to their mythology, the Maori dead descended to the underworld. Beyond, the Tasman Sea and the Pacific Ocean creased together and the world turned blue. It was a more elegant entrance than the limestone caves at the tip of the Peloponnese or the volcanic throat beneath Lake Avernus, Italy, the Old World portals. Orpheus the Greek, or Aeneas the Trojan — anyone who entered the Old World portals — was warned by shrieking goddesses and a noisy chorus of harpies and gorgons how easy it was to go in, how hard to come out. But at Cape Reinga, today at least, there was only a moderate salt wind, and nor did you have to descend through the portals at all. You could get there and get out again, mortality intact, for the underworld had heaved itself ashore and we were standing on it.

I'd persuaded Mike to come up to the Cape for two reasons. When mapping out a route for the Te Araroa trail in the 1990s and 2000s, I'd had a lot of New Zealand roll away under my boots. The continental variety of that landscape is well known, but I'd felt the changes underfoot as a slow and salutary revelation. I wanted to know more. I wanted to start in the north and seek out a longer history of the trail's best landscapes. But I was going to need help. I put out feelers to see who might explain the complex geology of Northland, and the word came back: Mike Isaac. I went to his Leigh house. There was a Morgan in the garage, a table of fossils beside the front door, and beer in the fridge. The man who owned the Morgan, who opened the front door and later the beer, was fit and feisty. He was bearded, as many geologists are, but it was a short trimmed beard and hardly softened an incisive face.

By reputation he didn't suffer fools, and by reputation he had an encyclopaedic grasp of New Zealand geology, gleaned both from fieldwork and running the QMAP project that laid out New Zealand's geology at a 1:250,000 scale. He was known as a maverick. There were witnesses to say Mike had been a boulder roller during his scientific stints in Antarctica, prising loose rocks and setting them bounding away for minutes at a time down the enormous runout slopes of the frozen continent, ricocheting off the nunataks, bouncing off the dolerite columns. Within his Leigh house I found ammonites in the bathroom and chalcedony quartz, sliced open and polished to display the growth rings.

In 2011, long before our first meeting at Leigh, Mike Isaac had come to Te Araroa's opening ceremony at Island Bay, a memorable Wellington day that began with the full pomp of an army band, and ended with a 5.3 earthquake. He was a walker who'd done the Camino de Santiago with his wife Maggie, and regularly walked the bush trails around his Leigh house. These were points of contact.

The Northland Allochthon — the Great Thon — was finally confirmed in the late 1980s, when a Geological Survey team that included Mike put it beyond doubt. Before that, from the late nineteenth century and for most of the twentieth, Northland's strange mix of rock strata had led geologists to question their field data and to contradict each other. It broke friendships, and it drove at least one of its early geologists to drink. For over 100 years the only description that seemed to fit the Far North was chaos.

— All that erosion in a subtropical climate, said Mike. All that rampant vegetation. It took a while for people to see that Northland was like the Swiss Alps.

In 1848, Arnold Escher, a geologist whose territory included the Glarus Alps in central Switzerland, walked over the Segnas Pass with the greatest geology luminary of that time, Sir Roderick Murchison.

Swiss geology allotted whole cantons to its distinguished scientists and they usually kept it within the family, passing on the jurisdiction and the knowledge father to son. The system was aristocratic, though in the Swiss way it should serve the public good. The social service of Arnold's father, Hans Conrad Escher, included engineering the Linth River to prevent the regular flooding of riverside agricultural land, but he also ventured deeper into the Alps, where geology became pure, and speculative, a goad to the imagination. Once, looking up at Mt Nollen, its upturning strata picked out by snow, Hans Conrad had been assailed by a strong vision of the earth's crust in motion, unfolding.

By the time Arnold Escher inherited his father's bailiwick, European and British geologists, Murchison pre-eminent among them, had developed a system of dating rock strata by index fossils. Arnold Escher had applied the system to the Glarus Alps, and his walk with Murchison was a chance to unburden himself of a vision more unsettling than anything that had happened on Mt Nollen. Folds within Swiss alpine strata were commonplace, including the most extreme form, the so-called recumbent folds that turned stratigraphy on its head. But as they walked, with the relevant strata exposed alongside and literally to hand, Escher showed Murchison where Jurassic rocks that were 200 million years old lay over Eocene and Miocene rocks that were only around 50 million years old. That inversion of the proper order, said Escher, stretched in his own jurisdiction over 10 miles to the Panix Pass, and he believed the inversion ran on, straddling the distance between the Bern and Glarus cantons, a distance of over 100 miles. It didn't look much like a recumbent fold.

Escher almost pleaded with Murchison to say he was wrong,

but the Scottish geologist could only confirm his colleague's work.

'I was convinced,' wrote Murchison to the London Geological Society in 1848, 'that M. Escher was correct in his delineation and mapping of the ground although he ingenuously urged me to try in every way to detect some error in his views, so fully was he aware of the monstrosity of the apparent inversion.

'It became necessary to admit that the strata had been inverted, not by frequent folds . . . but in one enormous overthrow.'

Murchison was struck by the overthrow's 'apparent uniformity, simplicity and grandeur'. Searching for a cause, he proposed granite ellipsoids rising from depth, huge buoyant intrusions, overturning as they rose huge tracts of overlying strata. Geologists had already located considerable masses of granite within the Alps, Murchison reasoned, and might there not be further upwellings 'hidden under unfathomable glaciers'?

Escher wasn't about to publish on such a controversial subject, and beyond his letter writing, conversation and field notes, news of the overthrow didn't travel. He was a cautious, even timid man. By the 1860s, exercised still by the problem but knowing that to propose it formally would be to invite questions of his own sanity, he settled on what became known as the Glarus Double Fold. It relied on the concept of a shrinking earth. The earth was cooling, contracting, and either side of that descending furrow of younger rock, the folds of older rock closed over, like a man descending some dank oubliette, closing the twin trapdoors over his head as he disappeared.

The inheritor of Escher's work, Albert Heim, endorsed the theory with a monograph published in 1878, six years after Escher's death. Heim's position as a professor of geology at Zurich, and his beautiful illustration of the phenomenon, gave it legitimacy and the double fold became a Church Triumphant over any thought of vast creeping slabs.

But outside of Switzerland, the concept of the allochthon

was already finding its way into geological literature — massifs with alien stratigraphy that arrived from elsewhere, overriding the autochthonous structures of the local country rock. In 1884, without even visiting the site, the French geologist Marcel Bertrand used his wider knowledge to reinterpret the Glarus Double Fold as a single thrust sheet sliding in from the south. It was a far simpler explanation. Jurassic strata had been thrust over Switzerland's basement rocks during a vast tectonic event some 30 million years in the past, but no one in his day knew where it had come from, or why.

In the same time period, on the other side of the planet, the once-mighty continent of Zealandia continued to subside into the sea. As it first rifted away from Gondwana 50 million years before, Zealandia had been 15 times the size of present-day New Zealand. It was the earth's seventh continent but by the Oligocene it was a shadow of that former self, stable yet passive, eroding into the sea and diminishing too by what geologists call thermal relaxation. The separation from Gondwana had stretched and thinned its lithosphere, and the heat every craton holds to itself, from convection of the mantle far below or by radioactive decay, dissipated. Growing colder, it lost buoyancy and gradually sank. As the Oligocene ended, Zealandia was a string of remnant islands, surrounded by submarine platforms of its original greywacke and by the products of its own long erosion — layers of yellow sandstones and, further out, the finer sedimentary material of grey mudstone. In the final years of Zealandia's demise, the shallow platforms gathered white limestones. For 50 million years the continent had danced slowly upon the world stage, but by the end of the Oligocene, Zealandia was taking her final curtain call, sunk down upon a surround of white and yellow and grey petticoats.

At around 23 million years ago, just offshore of the remnant continent, the rift between the Pacific and the Australian plates opened into a full subduction system. The Pacific Plate began its descent. Part upper mantle, part crust, a plate that was 70 kilometres thick now sought its accommodation within and below the Australian lithosphere. From this early Miocene beginning, whether by micro-tremors as its basalt changed at depth into the denser eclogite, or by the jolts of earthquakes large and small, it would remain a constant brutal shimmer of descending movement, and within that immensity the plate's accumulated skin of terrigenous and limestone strata would be stripped away from it and despatched westward.

Allochthon — from the Greek *allo* (other) and *chthon* (earth). A geological intrusion that originates and moves in from elsewhere, and that is not, to summon another of geology's more exalted and in this case directly opposing definitions, the autochthonous or stationary country rock.

Twenty-three million years ago, a vast allochthon was on the move towards the sunken bulk of Zealandia.

Mike was leaning against the white concrete plug of the Cape Reinga lighthouse and making plain with his hands the kind of movement and the kind of magnitude that might beggar belief.

— The original land mass is going down, he said, so you're making an accommodation space. The nappes are able to come in over the top, and if you start loading the earth's crust it starts to sink further. The nappes come in sequentially, this lot, that lot, probably about six major thrust sheets partly sliding under gravity and partly pushed by this classic subducting system. They haven't gone down the gurgler, they've been scraped off the top and are being pushed over the existing land mass. It's called obduction.

While we talked, Cape Reinga had become populous with visitors. They'd made a long and determined journey to this spot, and they faced into the wind and held up their iPads and smartphones in front of the view. They photographed the lighthouse. They photographed Mike and me as we stood there. They photographed the bristling yellow AA sign that stood beside the light. The sign pointed with ordinal authority to London, Sydney, Los Angeles, Tokyo and the South Pole and gave their distances in kilometres and nautical miles. I liked the sign. Te Araroa walkers gripped the sign before setting off, and the sign did a fine job of holding the world steady behind them, but any animation of the sign in geological time would see its pointers swing and its mileages turn over like an odometer. Tokyo was moving east, more than a metre during the March 2011 Tohoku earthquake alone, Los Angeles was headed north and compressing east, London inching north also, and the South Pole was subject to polar wander.

Mike Isaac leaned against the buttresses of the lighthouse, his fabric hat now clasped in his hands in deference to the wind, and he was talking in geological time.

— Back in the 1980s the critics of the allochthon theory were saying. 'You don't have enough time to do it,' but we were lucky with GPS and plate boundary measurements. They established Pacific Plate movement up to forty-seven millimetres a year. That's forty-seven metres in a thousand years and forty-seven kilometres in a million years. So we were able to show the rates we needed to make it all happen. The thing people never appreciate is that the length of geological time makes possible enormous change.

I asked Mike how far the allochthon had travelled. He was equivocal — 200 kilometres was a minimum, 500 a plausible, and 1000 a possible distance. The last of the thrust sheets had travelled the furthest distance and it was no longer the sedimentary layers on top of the plate but part of the plate itself — oceanic basalt. Not just Zealandia slow-dancing on stage, sinking down, then hoist back towards the light by the first outliers of the allochthon, but now the stage itself, ripped up and despatched. The basalt headland we stood on was hundreds of metres thick. Cape Reinga itself and the high-standing massifs that recur throughout Northland are all ocean plate — the so-called Tangihua basalts. The Tangihua basalts are up to 100 million years old and have been shunted over the top of younger rocks, an Antipodean recurrence of the monstrous overthrow that distressed the Swiss geologists and fuelled their rearguard actions against the vast creeping slabs.

The lobate forms of ocean-floor pillow lava were all around us in the cuttings as we walked back up the path from the lighthouse, and I paused to photograph a slab of sandstone engraved with a description of the Cape's geology.

> *Volcanic remains*
> *The dark-red rocks like these on the Cape are volcanic.*
> *They come from eruptions under water long ago.*
> *Movements in the earth's crust over millions of years*
> *have pushed these rocks above sea level.*

— That's both true and underwhelming, said Mike. If I was writing it I'd probably leave out the word 'allochthon' too, because it would numb the senses. But I might have difficulty avoiding the words 'subduction' and 'obduction'.

Fourteen kilometres south from the Cape we turned off the state highway and drove to the road end above Te Paki Stream, engaged four-wheel drive on the Mitsubishi and followed the stream through to Ninety Mile Beach.

At Scott Point we explored towering piles of pillow lava washed clean by the sea, then turned and started down the beach. The long low dunes on the left and the surf to the right converged ahead into salty haze. Utes scoped the beach, GPSing the low surf lines that suggested a hole offshore. Big snapper came to feed in the holes.

The Ninety Mile Beach Snapper Contest was soon to begin, and on competition days it was an advantage to know the location of the holes, drive straight to them, claim them, surfcast into them.

Halfway between Scott Point and the Bluff, Matapia Island stood offshore, ruffled by surf, bright as a button, with a hole in the middle.

— The island is an oddity, said Mike. Why is it there? In the middle of the beach, this funny little pin sticking up.

Earlier geologists stood on the beach and mapped the island as Tangihua basalt, but during his work in Northland, Mike had a feeling they were wrong. He got a chopper pilot who owed him a favour to drop him and a mate onto the island. They prowled the island, sidestepped a bull fur seal, tapped the island with a geological hammer and found only indurated sandstone and, astonishingly, a fossil of reef coral. Matapia was a scrap of the rocks laid down in warm shallow seas, after the allochthon had slid into place.

We drove on to the Bluff, an exposed relic of old Zealandia, collected salt horn from its small wild pastures, then kept on down the beach. In front of us, black-backed gulls took tuatua aloft and dropped them to crack their casings on the beach below. The gulls took the tuatua high, but not so high that the neighbouring gull might move in and snatch the prize before the descending one could land. The regular spacing of the gulls along the beach suggested they were happy with the role of either procurer or thief, and they moved towards each other, or retreated, seeking advantage for either role.

The Aupouri sands don't hold much to interest a hard-rock geologist, and it's left to the archaeologists to gather charcoal and giant toheroa shells in the middens, date them, and suggest that Maori all cleared off around the end of the sixteenth century, reasons unknown. But geologists have had one suggestion. In places the Aupouri dunes are mantled by gravel that shouldn't, by any usual motive force, be there. Gravels gather as colluvium at the foot of steep rocky slopes, but there are no rocky slopes above the Aupouri gravels. Gravels may be brought in by water, but there is no evidence of streams. That leaves the ocean. Geologists confirm the source of the gravels is greywacke and basalt chunks, broken off the headlands and rounded into cobbles and pebbles within the surging surf below. The surf zone is where they should stay, too heavy to make the beach.

A 2004 study that investigated Henderson Bay on the eastern side of the peninsula spoke in the classically restrained voice of a geological paper: 'A fluvial or colluvial origin for these deposits is discounted . . . due to their isolation from streams and hillslopes. We also discount the possibility that these deposits are the product of aeolian transport on the basis of clast size (up to 67 millimetres) and elevation. We propose they were swept there by tsunami.'

The probable origin of the tsunami, that paper supposed, was the Healy Caldera, a submarine volcano some 520 kilometres off New Zealand's eastern shoreline whose eruption and subsequent structural collapse left a sunken throat three or four kilometres wide. The signature sea-rafted Loisels pumice that's distributed far and wide across the eastern coast of New Zealand is thought to have come from that eruption, and the 2004 study found the Loisels pumice in a buried layer behind the mantling gravel at Henderson Bay. It led the researchers to suggest a date for the tsunami of AD 1450, and a height of 32 metres, comparable to the tsunami that killed 280,000 on coastlines around the Indian Ocean on Boxing Day 2004, or the tsunami of 11 March 2011 that killed 16,000 on Japan's coast.

Research since 2004 has not confirmed Healy Caldera as the source. Many geologists believe only a large subduction earthquake along the Kermadec Trench could generate a 32-metre wave. Whatever the source, there's acceptance that at least one large tsunami swept the peninsula in the fifteenth century, and that there might have been other waves earlier, and later.

We came off the beach at Waipapakauri and turned back onto State Highway 1. The Gum Diggers Park was a short distance off the highway and we took dirt paths, then the boardwalk and steps, down to a kauri trunk preserved in peat, and now excavated so we could reach down to lay hands on the strange upper surface of the fallen giant. That surface was abraded flat along the entire length of its trunk, as if run through a buzz saw.

The park's meandering paths showed the remnant holes and tools of the gum diggers. It had reconstructions of their shanties, and photographs of men working their waterlogged industry, but hints of an ancient violence haunted the site. Kauri that once colonnaded large parts of the peninsula didn't die their usual 1000-year-old deaths here and rot out still standing. They died flattened in the peat, and where they've been exhumed by bulldozers for commercial use they're often found flattened in a single direction.

The kauri forests were already gone from the peninsula when the AD 1450 tsunami swept though, but geology is repetition: what happens once will have happened before, and will happen again. I left the park with visions of staunch old forest giants toppled by some tsunami out of their own time, abraded first by the onrushing sea of debris, then laid flat.

Back in Kaitaia, we drove to Peter Griffiths' house at Takahue and dined that night on salt horn fried in butter and fresh tuatua. Next day we drove up the Gumfields Road into the high plateau behind Ahipara, another chunk of the allochthon's Tangihua basalt. Kauri forests once grew here too, and left a bonanza of gum, dug out last century by Maori and by Yugoslav immigrants.

White windblown sand was creeping in long fingers across the old gumfield and we looked north over the long stretch of dunes we'd traversed the day before.

— It's quite dramatic the way New Zealand runs out as you go north, said Mike. It's the tail of Maui's fish. Just low-lying dunes and there's nothing older than a few million years in them. The

Tangihua basalts and the Mount Camel basement over there poke up through them like islands. We've only just arrived back in NZ now, and there's just islands tied on by the dunes. It's a long way. People think, 'Oh Kaitaia, that's in the Far North.' It is in the Far North, but you can go a hell of a lot further on that long skinny finger of sand.

We drove south down State Highway 1, into New Zealand, and into the allochthon. It stretched from Cape Reinga to Silverdale. Its leading edge had overshot Northland and rumbled on up to 100 kilometres into the Tasman Sea. Mike and his colleague Rick Herzer had estimated the bulk of it at 100,000 cubic kilometres. That was as big, measuring from shoreline to summit, as 36 Coromandel Peninsulas, and the obvious question was why it took New Zealand geologists so long to recognise it.

Mike was a connoisseur of Northland's geological history. He knew the mindset imposed by the Old Geology. The Old Geology had refined its index fossils, and knew the ages of its rock strata. The Old Geology knew the earth moved. It moved up, and it moved down, but it didn't move far, so if the stratigraphic column didn't make sense, Old Geology turned to the somewhat occult concepts the Swiss had employed on their alps — hidden folds, or hypothetical faults, or broken outcrop patterns. Mike didn't hold it against the nineteenth-century geologists that the allochthon eluded them. The name didn't even exist back then, nor any credible mechanism for such enormous shifts. What he did look for was accuracy of observation.

Within the shade of the Old Geology, the ones who coped best with evidences of maximum movement, he'd decided, were the ones with minimal formal training. Nineteenth-century New Zealand needed to search out minerals and coalfields. Reliable geological assessments were a critical part of such prospecting and in 1865 the government appointed Dr James Hector as the first director of the New Zealand Geological Survey. In 1877 Hector came north to the coalfield at Kawakawa. His report placed the Kawakawa coal correctly as Eocene, but the most obvious of the layers above the coal was Whangai shale. The shale had ammonite and inoceramus fossils. It was irredeemably Cretaceous, which made it older than the coal by some 30 million years. The Whangai shale didn't fit, and the university-trained Hector simply left that top layer out of his 1877 report. The Geological Survey's field geologist, Alexander McKay, came through in 1884. He was self-educated. His beautiful map and cross section put the Whangai shale above the coal.

— McKay had no explanation for that, but it was accurate, said Mike. But Hector said, 'No, that can't be the case. I know they're older than the coal, they have to be underneath it.' But McKay could map it all and work it all out just from his field geology. If you look at it nowadays you know McKay's right — there's a big thrust fault separates that and that. McKay's instinct as an observer was to record what he could see in the field, whereas others were saying, 'No no no, that can't be right. I know the age of the units and that doesn't work.' That lovely contrast.

— And most people later on, until well into the 1950s and 60s, followed the Hector approach: we know these Whangai rocks are older and therefore they must underlie the coal. The whole dispute went through to a guy called Hartley Ferrar, who'd been to Antarctica with Scott in 1903. He'd joined the New Zealand Geological Survey later and was sent up to map Northland. He was a classically trained Cambridge-educated geologist, and the response was, 'Hang on a minute. We know they're old and we know the coal is younger, so the coal cannot be underneath those rocks, it must be on top somewhere.'

Various geological reports followed that pushed the rock layers around Kawakawa and other coalfields down into the Cretaceous, or up into the Oligocene, and while no layer could be pushed beyond boundaries established by its index fossils, still, confusion reigned on what was collectively called the Onerahi Formation, and that was before anyone even considered the big blunt problem of the Tangihua basalts.

In 1943–44 Larry Harrington, a 20-year-old Auckland University student, went north to explore a wide chunk of upstanding Tangihua basalt, the Whirinaki Range just south of the Hokianga. He was faced with three possibilities. The range could be a high-angle fault block pushed up through the sedimentary layers of the Onerahi Formation, or it could be an anticlinal dome exposed as the overlying Onerahi Formation simply eroded away around it, or it might have erupted through the formation.

His fieldwork yielded only objections to each alternative. If the range had emerged as a fault block, then the fault had to run all the way around the 40-kilometre diameter of the range. Such high-angle enclosing faults were rare. If the range was a dome uncovered by erosion of successive overlying layers of the Onerahi Formation, then the emergent dome should present, as the layers fell away around it, a basalt bullseye to the Onerahi Formation's eroded roundel. There was no bullseye pattern. If the range had erupted through the Onerahi Formation, then the basalt lava would be seen to interlock intimately with the surrounding sedimentary rocks. There was no interlock.

Harrington's paper posed the problems, and no solutions. That same year, the most intuitive geologist New Zealand has ever produced, Harold Wellman, visited the north briefly, gathering samples that he sent on to the Geological Survey's paleontologist Harold Finlay. His covering letter to Finlay said, 'Having seen some of the geology from Kamo to Kawakawa some kind of overthrust appears probable.'

— If Harrington and Wellman had sat down for a beer together the allochthon would have crystallised forty years earlier than it did, said Mike. They were contemporaries. They were right on the cusp of working it out.

But Wellman was already preoccupied with refining his theory of massive movement on the Alpine Fault. He didn't pursue his Northland intuition, and Harrington joined the Geological Survey but was sent off to work on other things.

That was that, until Mike was cleaning out departmental files in the Geological Survey's Auckland office in the mid-1980s and came across a note dated August 1955. By then, Harrington had seen evidences in Cyprus of the big bold movement within the Troodos Mountains. Tall chunks of what was believed to be seabed had been uplifted almost intact. Harrington put in a two-page note to his employer. It was sufficiently heretical to be filed and forgotten until Mike Isaac came across it.

Harrington's note recalled the 'puzzling stratigraphic and structural relations of Tangihua and Onerahi rocks' outlined in his 1944 thesis and stated: 'The stratigraphical and structural difficulties in interpreting Whirinaki Range disappear . . . if it is envisaged that the range is a nappe separated from underlying Onerahi rocks by a slide.'

He surmised further that such an explanation for the Whirinaki Range could be extended to all the Tangihua basalts. They could be understood as 'portions of a formerly continuous sheet of Cretaceous or pre-Cretaceous igneous rocks that have been thrust for a minimum distance of tens of miles . . .'

Harrington's heresy deepened further as he proposed that not just the Tangihua basalts, but the whole 'nappe facies', the entire suite of sandstones, shales, limestones and mudstones underlying the basalt, might be allochthonous.

Mike Isaac read the note with deep delight.

— What that note said was that years ago I, Larry Harrington, did a Master's thesis at Auckland University and I didn't have a clue what I was looking at but now I've been to Troodos and I know

now what my Master's thesis was. I know now the Tangihuas are allochthonous, that they're bounded by a thrust fault underneath, they bear no relation to rocks anywhere near them or around them, and in fact that leads me to question whether any of the Northland geology is in place. That was Larry Harrington. He was rubbished, at the time, but he had it in a nutshell.

In the 1950s, the note lay forgotten in the files, and the interpretation of the Onerahi Formation continued in confusion. The science moved from large index fossils to the progressive expansion of microfauna analysis, which yielded more precise information, more locally relevant, but that didn't help. By the 1960s, the upper sandstone and shale layers of the Onerahi Formation had yielded microfauna ranging from Cretaceous to Miocene in age — a range of over 50 million years, and on some of the outcrops you could pick up that variation within a few square metres. Something had milled and mixed the rock, and the geologists sought to solve the so-called 'haphazard lithologies', 'exotic rock types' and 'chaotic development' of Northland's rocks by branding it Onerahi Chaos-Breccia. Where had it come from? The best bet was slumping, the origins of the slump unknown.

We drove on through the Maungataniwha Range south of Kaitaia and Mike gestured at the grey ramparts that ran alongside.

— These are massive basalts, said Mike. The north has about a dozen big lumps of Tangihua basalt and Maungataniwha is the biggest. They're rootless. If you do a gravity survey over these you might expect there's a lot of dense basalt material underneath. But the gravity surveys show that's not the case. The gravity anomalies show the high ranges are just a superficial skin. There's a thrust fault beneath them and they've been pushed in over the sedimentary rocks.

We turned off the highway at the Mangamuka Bridge settlement, driving up to Broadwood Road to look for the thrust fault that separated a Tangihua basalt from its softer underlay, then to work our way on down-section. The road began to snake around the base of a basalt Maungataniwha outcrop and we pulled into a side road.

— We're within fifteen metres of the transition, said Mike. The range front is quite steep and you could follow that around for thirty or forty kilometres until it hits the sea, then underneath it is this soft Cretaceous hill country.

The rounded ridges and hummocks of a Northland hill-country farm stretched away below the road and we followed the Broadwood Road down to stop on the grass verge and stare at a crumbling road-cut. Wild daisies nodded at its foot, and it sloped upwards to a lip of kanuka and crooked fence posts.

— If I was still in GNS Science now I couldn't do this, said Mike. Just stop by a roadside outcrop, and get out. I'd have to put on a high-vis jacket, and that jacket must be zipped up before I leave the vehicle. And we'd have to put out traffic cones along the road.

— The bright orange high-vis vest is particularly important to signal your presence, I said, when crossing a paddock of bulls.

— That's right, said Mike. The bulls have got to know where you are.

We laid hands on the brittle Whangai shale. Top to bottom the road-cut was milled and smashed, and some of its larger rocks had been rotated, the pressure shadows flung outward like some slow lithic Catherine wheel.

— If the rocks were sandstone and relatively soft then they didn't deform like this, but if it's a shale and has a bit of resistance then it's just crushed and smashed, said Mike. You can see kernels of relatively undeformed rock still in it, but the whole thing has got that crushed and sheared look to it. Beautiful. You can't believe that other geologists looked at this for so many years and didn't twig to it. If this was a solitary occurrence you might try to think of another, an extraordinary explanation for it, but it's a pattern you see throughout Northland. Look, you're beneath that big high basalt range front over there.

He pointed to the basalt ramparts we'd just left, then opened his hands to present the next layer down, the smashed bank in front of us.

— You can practically see the contact. And then you find that repeated again and again throughout Northland. What would you think?

We drove down the Mangamuka Road, and dropped to the allochthon's next level. The limestone ran alongside us as a grey cliff with overhanging bush and creepers and Mike gestured up at it.

— Some countries have been lucky enough to have a glacier come through. If Northland was glaciated and all this nonsense had been scraped off the top of it, everything would be obvious. The Northland Allochthon story would have been plain because you would have seen things in an inverted order, but it's obscured by the low relief, the vegetation, the severe weathering. It became inference from bits and pieces.

We got back on the highway then turned off onto the Omahuta Forest roads. Mike wanted to find what he called 'the unctuous grease' — hemipelagic ocean floor muds, distinctively coloured, that smoothed the progress of the allochthon's first incoming layers of mudstone and limestone and had eased each of its subsequent thrust faults.

— There's the red and green!

Thick bands of pelagic colour slanted across the forest road-cut. Dried at its surface by sun and wind, the mudstone was brittle enough that we could pick away at its distinctive laminations. Unlike the milled and rotated fragments of the Whangai shale, the mud bore the evidences of vast sliding pressure in striations of planar shear and polish.

— That's a great outcrop, well worth the trip into the forest.

Mike stepped down from his examination of the road-cut and sank into hemipelagic mush in the gutter.

— Hmmmnnnn. I like soft rock, he said, flicking a Teva clean of the mud, but not this soft.

We stood at a lookout on the Omahuta Road overlooking a wide valley. We'd come down-sequence through the allochthon from top to bottom and here it fell away at our feet into an eroded valley, then rose again as a single Northland landscape, but Mike was looking into that landscape to separate the layers.

The Omahuta Road ran through the valley below, and a half-round hay barn and galvanised sheds stood in a green field — under that pasture, the mudstone layers. A gradually ascending farm road led away through fields and pine windbreaks, then was lost to sight as it climbed a bush-clad ridge — under the bush, the limestone layer. Beyond the ridge, the land dropped away again unseen, but as we stood on our lookout, maps to hand, we could predict the same soft green fields we'd seen on Broadwood Road, sown for pasture, top-dressed, but always quietly eroding into the rivers and harbours of Northland — the Cretaceous sandstone and shale. The allochthon had been converted to sparsely settled farmland across most of its reach, and then the Maungataniwha Range rose as a shadow in the west, a telco tower pinned to its highest peak. The basalt.

— These things, in reverse order of age, dipping away from you, said Mike.

We drove to the nearby Apple Dam campsite, parked, and set off walking Te Araroa's route through Omahuta Forest.

The forest was dense with manuka, and we passed hives brought in on 4WD utes. Flight paths of foraging bees converged on the hives, and we buttoned up to walk through them. Young kauri rickers stood beside the track. The road was little used, with a grassy median. It began to slope downhill, a good sign, for now we sought to go under the allochthon and find the original Zealandia, the country rock. We came to a ford across a stream, and a DOC sign warned that any flow of water over the ford signalled impassable torrents on the streams and rivers below. Walkers confronted with such a spill should turn back. The ford was dry and we kept on.

In 2007 a group of us had come along these same 4WD roads, seeking a simple route for Te Araroa from the grassy ridge roads down through the bush to the Mangapukahukahu Stream. We'd begun the bush-bash down. It had got steep, then steeper yet, but by then we'd been too far down to turn back. Though the protruding vegetation gave some assistance, shattered rock and slippery humus made the ongoing descents dangerous. We'd begun to hand-over-hand down precipitous faces. We'd posted helpers on ledges, to give advice on footholds for those still climbing down, and to steady them against toppling over the next drop-off. Falling rocks were a risk. No one should descend until those helpers below were out of harm's way. That was the rule when, crouched above the two Alpine Club trampers who were still securing their position out of harm's way, I'd dislodged a boulder the size of my head. It went bounding away, leaping and spinning. I shouted the warning. The rock had crashed on, ricocheting off the trees and then everything had gone quiet and a voice came floating back.

— Nancy's been hit. She's bleeding from the head.

By luck alone, the rock had struck only a glancing blow, and after some on-the-spot first aid, a dazed Nancy finished the descent and would walk out down the Waipapa River.

Some months later we'd found an easier route down, and seven years on, as Mike and I came through, a DOC sign was now in place to confirm that route. We followed orange triangles on waratahs to an old skid site, then descended through young pines towards the Mangapukahukahu Stream. Mike broke through the trees ahead of me. He jumped down onto the pebbled bed of the stream and I heard him yell, then splash through to the far bank where he was already reconnoitring the rock.

— Chapple's luck persists. His Te Araroa track happens to come through on this beautifully exposed geological contact! You've fluked it. You are without doubt a tin-arsed bastard and know far more geology than you ever let on. I got my boots wet because of it.

I jumped down onto the stream bed and looked about for the revelation.

— Bingo! said Mike from the far side. I know where we are. I know what we're looking at.

It took a while but then I saw it too, a slice of old Zealandia, biding its time within the Omahuta Forest. A peeping surface of basement greywacke, above that four metres of limestone and shelly sandstone, above that again a 15-metre cliff of grey mudstone.

The strata stood there in mute testimony. We saw old Zealandia's greywacke surface, once profuse with vegetation, once populous with dinosaurs, then the sinking of it and a shallow transgressing sea that favoured formation of limestone and sandstone. We picked at shell fragments broken by ancient storms. We rubbed embedded oyster shells, catseyes, and the delicate filigree of a bryozoan. And then as the greywacke and its carbonate layer sank still further we saw the light fade and the mud take over.

— You can see, said Mike, waving a hand upward to encompass the height of the mudstone cliff, the water is deepening. There's no more coarse sediment getting in. It's terrigenous mud, the Mangapa Mudstone. This is outer shelf material now. We're right out the back. This is a beautiful logical sequence. All laid down quietly. It's not shattered, it's not sheared. This formed on the seabed, around fifteen million years before the allochthon came in over the top.

Apple Dam was a pretty little campsite in the Omahuta Forest, primitive, but it had everything you needed. The track led past a long-drop toilet camouflaged from obvious view by punga trunks, then past a concrete water tank with a chromium tap, down to flat land that edged off into reeds and lily pads. Beyond that the dam deepened into a reservoir. Beehives had been placed here too, but on a higher terrace and the bees flighted in above our heads. As evening came on, the foraging path diminished and a light rain began to fall. Some previous camper had collected dry manuka sticks, and as Mike put on his coat and started the cooking, I got a fire burning bright in the drizzle, then broke out my WhisperLite to cook the potatoes.

I'd done it a hundred times before — pumped up the pressure, opened the valve fractionally to admit fuel into the priming cup, lit the cup to heat the fuel line sufficient to vaporise the fuel, then opened the valve fully and watched the burner head flare, then settled into the whispering blue incandescence that gives the stove its name.

On this night, in drizzling rain and darkness, it went wrong. The O-ring from the fuel bottle leaked its volatile fluid, and a nasty little inferno erupted around the bottle. I stamped it, and whatever patch I stamped out reignited as soon as my boots lifted. I had a moment of thinking my crotch was now poised above a bottle that could explode and my mouth opened on a cry — 'What now?' — when suddenly an allochthon came flying in from the east.

A tea towel. Mike had picked it out of his tent and thrown it across. Sudden brown burn marks spread across the cloth and wisps of smoke drifted up, but he finished an efficient smother by soaking the smoking tea towel with squirts from a water bottle.

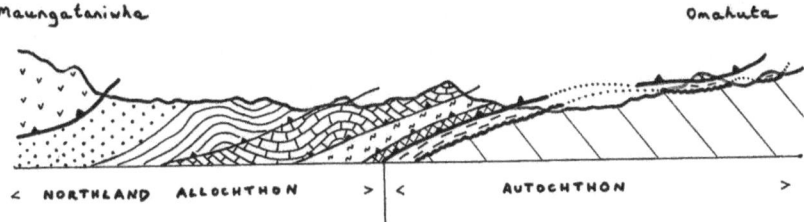

Later, after we'd returned to Auckland, Mike would send me a map of the strata that lay both in front of and behind us as we stood on the lookout on Omahuta Road.

In that same email came a gentle hint on how to find a route down into the Mangapukahukahu Stream without almost killing a member of the scouting party.

'Have a look at the topo map. If the land is near-flat or gently sloping it's Mangapa Mudstone, but I'd say you wandered off the soft and gentle terrain of the Mangapa Mudstone and into steep country of the erosion-resistant Caples Terrane basement.'

I agreed with that. Back in 2007, Nancy had been hit by a rock that had waited over 200 million years to begin its bounding flight to the stream below, drawing blood along the way.

The talk was of conspiracy. Who salted the drill cores on Maungaparerua, and why?

— Hi guys, said Mike to someone just beyond my left shoulder.

We were parked up in the hills above Kerikeri, and it was raining. I turned to see who it was. Fourteen big black heads hung over the fence line. Mist and rain closed down the view and made the big heads, the big eyes, the big ears all the bigger. A bovine audience jostled each other slightly for position, but remained entranced.

— It happens time and again, said Mike, and each time it's utterly charming.

We kept talking but it only deepened the mystery of how a

reputable New Zealand company had ever had its shares inflated by core samples that were, to use the polite term, contaminated. Who did it? It was an old scandal out of the 1970s but remained unsolved and Mike was giving his thoughts on the matter. I turned to the cows.

— Are you listening to this, guys?

— That one there might have done it, said Mike. The one with the white dot.

Maungaparerua, a modest hill of 240 metres, was one of my favoured places on Te Araroa's Northland route. After days of tramping from the western coast through the Herekino, Raetea, Omahuta then the Puketi forests, Maungaparerua offered the first glimpse of the jigsaw eastern coast. It overlooked the Bay of Islands and the distant shelterbelts and orchards around Kerikeri township.

The hill is an old volcanic dome, and its long-ago heat and bubbling springs have produced a rare white kaolin clay. The clay attracted Consolidated Brick and Pipe Investments Ltd (CB&P), owners of Crown Lynn pottery, but the same hydrothermal system that produced the clay could also produce metallic sulphides, and in 1969 through 1970 the company began drilling to establish what metals might be present. It sent the drill-core samples to London for an assay by the reputable precious metals firm Johnson Matthey.

On Friday, 6 November 1970, CB&P issued 300,000 ordinary shares at a premium for cash in a private placement. If you wondered why, the only clue was a bland press release that explained it was 'to provide funds for the company's work capital programme'.

The issue of such shares to a phalanx of unidentified investors, at a confidential discounted price, caused no comment from the financial journalists of the day. New Zealand law set no restriction on the discounts that might operate around private placements, and nor did it impose, as the Americans did, any restriction on immediate resale of such shares. The financial journalists didn't respond, but the Auckland sharemarket did. CB&P's share price fell almost a dollar to $1.69.

And then whatever ill feeling might have existed vanished anyway as everyone's boats rose. CB&P's share price began to escalate as soon as the exchange opened on Tuesday, 9 November, and bounded upwards towards a high of $4.15. The most significant boom in years was under way, and after a few hours of it, the Stock Exchange demanded to know why. CB&P's directors informed the exchange that an assay by Johnson Matthey had affirmed economic amounts of platinum at Maungaparerua.

Next day the *New Zealand Herald* ran three front-page stories. One canvassed the rapture of Kerikeri locals. One quoted the London assay. The third, under a four-column headline 'Deposit to Make a Director Smile', quoted CB&P director Tom Clark. The London assay had still to be confirmed, said Clark, 'But you can see that I am smiling. We have got the stuff. The question is how and whether we can get it out economically.'

Certain people — geologists — thought CB&P might *not* have the stuff. Where were the group metals that usually occur as part of the platinum suite: the rhodium, the palladium, the osmium and the iridium? Where was the nickel and copper? A fossil hydrothermal system like Maungaparerua would produce silver in trace amounts almost certainly, and if you were lucky some gold. It might produce platinum in trace amounts but almost certainly nothing more. The white metal is rare because it needs huge grunt to make a lode. It needs great masses of slowly cooling igneous rocks to draw in mineralised water and condense the platinum minerals within those deep layers. Fossil hydrothermal systems like Maungaparerua were superficial bubble and squeak, by comparison — interesting, but not the kind of forceful crucible that produces platinum.

Maybe some of that geological wisdom escaped into the public realm, or possibly the buck fever of 9 November simply cooled, for the same day the *Herald* trumpeted confidence with its three front-page stories, the share price fell to $3.25. It stabilised around $3.43 at the end of November, then dropped in the months that followed to around $2.40, and not without reason. Despite the encouraging assay results from London, local surveys couldn't find the mineral-bearing rocks.

Because of that impasse, CB&P called in the Department of Scientific and Industrial Research to report on rock types at Maungaparerua, and to provide referee assays. That work began in May 1971. CB&P provided 17 samples it wanted analysed, including remnant samples of the powdered samples sent to Johnson Matthey. The London assayers had used plasma emission spectrography to detect platinum in their samples. The DSIR had a machine as good, but the 17 samples yielded either no platinum or platinum only slightly above standard background rates. That was disturbing news. The DSIR reviewed its procedures and rechecked the results before reporting to CB&P that the samples didn't suggest economic quantities of platinum, and that the London assays were in error.

The DSIR was still at work in February 1972 when a new assayer, South Africa's Impala Platinum, reported back high, though sometimes erratic, platinum values from Maungaparerua samples. For the first time, the DSIR also got high readings. Isolate heavy residues, excite them with high-temperature plasma, capture their electromagnetic radiation and separate out from the noise platinum's distinctive spectral signature. That was the assayers' method of detecting platinum, and that should have been the end of it, but not this time. DSIR scientists now had a platinum-rich sample to hand, and they dumped sophisticated analysis in favour of No. 8 wire pragmatism. They put the sample's non-soluble residues into a paper box, tipped the box slightly and tapped the bottom until the larger particles rolled away. That left the fines. They put the fines under a standard microscope. Bits of it gleamed.

They sharpened a matchstick down to a single fibre, wet it with saliva and picked out the gleaming specks.

The specks were refined platinum. Their size, and the striations on their surface, were consistent with filings from a 0.7 mm diameter platinum wire. Thin platinum wire bends readily. If you decided to file it, you'd need to hold it very close to the file. Tiny slivers of what the DSIR investigators called 'collateral' were entangled with the filings. Whitish slivers. The scientists puzzled over them, then hooted with laughter, and swung the full weight of their optical petrology onto a triumphant description of those slivers.

Refractive index slightly less than 1.55
Anisotropic with a low birefringence
Mottled extinction pattern with parallel or sub-parallel extinction
Positive sign of elongation

In short, fingernail filings.

In April 1972, even while a sceptical DSIR was hard at work proving that the assays received from overseas were a crock, CB&P announced, on the basis of the reports it had from Johnson Matthey, that its platinum prospect at Maungaparerua was worth $33.7 million. This time the sharemarket barely responded. The CB&P shares gained 20 cents to $2.70 then fell back to $2.32 at the end of the month. Soon after, when the DSIR reported that the platinum in its sample was too pure to occur in nature, CB&P called in the police. Another batch of assayed samples came in from Nippon Steel, Tokyo, with platinum values described by the DSIR as 'enough to set off a mining exploration or sharemarket stampede' but by now the samples fell under police custody. Examination showed the same unnatural purity, the same entanglement of guilty fingernails. The DSIR investigation had now turned from referee assaying to forensics done on behalf of the police. On 5 October 1972, CB&P announced they no longer placed any economic value on the platinum prospect at Maungaparerua. At the same time, police confirmed a fraud investigation was under way. CB&P shares dropped back to $1.72.

The DSIR duly identified the interval when the fraud had taken place. It occurred after the final grinding of specific cores into a preparatory powder, and before the dispatch of such powdered samples overseas. Narrow that window might have been, but in July 1974 the police advised the DSIR that no prosecution would be made.

Forty years on, when I asked to see the fraud file, the police turned over every archive, as they were bound to do under the Official Information Act, and reported that all records of the scandal were gone. So much for one of the biggest mining scams in New Zealand history. In 1974, Consolidated Brick and Pipe Investments, a company beloved of the country's top politicians, changed its name to Ceramco.

We stared into the rain.

— Why even analyse the cores for platinum? said Mike. It's out of the bounds of reasonable prospecting. Platinum likes ultrabasic and basic environments — rocks with low silica. To look in a hydrothermal deposit was a leap beyond the credible. Yet when they looked for platinum they found it . . .

I turned to the cows.

— I think we're right to blame the one with the white spot.

— He'll do, he's different. In fact, is that a pink star on him or a yellow star on his coat as well? We'll take him anyway.

— That's a yellow plastic platinum rort tag he's got on him.

— You're in trouble, one-seven-six.

We looked out at the dome. A telco tower stood on the summit. Maungaparerua might not have been a platinum bonanza, but it did have kaolin clay. The hill is part of Landcorp's Puketotara farm, and in the course of signing up the agreement on Te Araroa's through route in 2010, we'd accepted a clause that the track could be closed for a week or two, if the owners so wanted, to mine the clay.

The clay was interesting in itself. Take a rhyolite volcano from the Miocene, allow to degrade slowly into a softish dome. After 15 million years, open up a few entirely different volcanic vents, and allow a valley-hugging basalt to swirl around the dome. The thicker the basalt the better, since its heat will drive hydrothermal systems that turn the already degraded rhyolitic rock, rich in silica, into white clay. Importantly, the surrounding basalt will also enclose the clay and stop any rapid erosion. Allow to cool and await discovery.

Over 1000 years ago, Chinese artisans discovered and mined a similarly modest hill in Jiangxi Province. Gaoling village, at the foot of the hill, would later give its name to kaolin clay. When fired at high temperatures the clay yielded a hard, white and translucent pottery that intrigued first the Chinese emperors and later the world. Marco Polo, during his thirteenth-century visit to China, termed it 'porcelain'. Jingdezhen, the city downstream from Gaoling, would later produce the prized porcelain, also lesser-quality whiteware, on an industrial scale, export it, and for centuries preserve to itself the secret of the clay and the methods of firing. Jingdezhen's distinctive ceramics became so ubiquitous that the country of origin lent its name to the product itself — China.

Once kaolin's chemical properties were known, the West hunted down its own sources. By the time the owners of Crown Lynn pottery got to Maungaparerua the clay was no longer a rarity on the world market. Still, it was a good domestic source. The basalt flows that cupped the clay were 30 metres thick and it wasn't going anywhere fast. Well, not that fast.

— The water that's flowing down that gutter is white, said Mike. It's carrying clay in very fine suspension.

We drove down to Kerikeri township. I'd rung ahead to sculptor Chris Booth, and he and his partner Anne-Marie had invited us for lunch. When the time was right, I was going to put a question to Chris. It was a tricky one to put to a busy sculptor, who now worked internationally on big commissions, but he was from Northland, he loved Northland, he worked on instinct and you never could tell.

Twenty years ago, when I and a few enthusiasts had put in a walking route between Kerikeri and Waitangi, we'd hoped that would start the resolute effort for a New Zealand-long track. It was Te Araroa's first linking track, it seemed symbolic, and I'd gone to Chris and said I'd like to mark its opening with a plaque and a sculpture. The money we could offer him was tiny.

Chris didn't hesitate. He proposed a cairn.

I designed a plaque to fit the occasion, sifting through New Zealand poetry for a quote that might sum up the spirit of a new trail, and came to a couplet from Rex Fairburn's poem 'To a Friend in the Wilderness'.

I could be happy, in blue and fortunate weather,
roaming the country that lies between you and the sun

I had it cast in bronze, and Chris used a diamond cutter to incise a lump of carefully chosen basalt, cold-chiselled a rectangle, drilled holes to accommodate the lugs on the back of the plaque, and epoxied it into the stone.

He drilled out over a hundred smaller rocks, threading them onto a stainless steel cable, winding them into a spiral, and clamping it tight with small-gauge stainless steel wire —Te Araroa's scoria cairn. Aside from its single drill-hole, every rock had as many tiny holes and brittle cells as a fossil sponge, and we figured when Te Araroa got going and its walkers came through they might dress the cairn with a sprig of greenery.

The plaque lasted a couple of years, then intruders entered the forest, smashed the rock and levered the plaque free. They used the bull-bars on their 4WD to shunt the cairn until the steel helical cable snapped at the base, the cairn slumped back over a steep bank, and the rocks dripped off their binding. Twenty years on from that first opening, Te Araroa walkers were coming through by the hundred, but the symbols of the trail's very first opening on Te Puke Forest Road were no more than a flaccid sock of red rocks draped down a bank and enclosed by bush.

We sat at an outside table with a spread of bread and olive oil, cheese, tomato and rocket, and Mike gestured out across the grasses that stretched between the house and the river to something that caught his eye.

— That's a nice fluted basalt.

A big grey clam sitting in the meadow. So big you'd need a crane and a truck to shift it. It struck me then that Chris was a basalt man. He'd grown up on the basalt flows where we sat. He'd bathed where the basalt held the Kerikeri River in its clasp, and all around had been a playground of basalt boulders, or the cups and shafts the river had hollowed out of the wider basalt flows. To enter his adobe house, we'd taken our shoes off at the door, and stepped across a threshold of basalt. Anne-Marie had prepared lunch on a basalt bench.

Basalt had given Chris a destiny. He'd torn his shoulders with the big rocks more than once, and been tested by them. In 1987 he'd stood on the wild coast south of Matauri Bay with elders of Ngati Kura as they lifted the tapu, and told the boulders they were going to Auckland. Two- and three-tonne basalt boulders packed tight by centuries of storms, and Chris attached cleats when the tide was low, bound 44-gallon drums to the cleats and waited for the lift of an incoming tide. Or lifted them by a Tirfor winch on the bow of an aluminium tender, counterweighting the basalt plumb bob in front by pouring water into a 44-gallon drum lashed to the stern. The big black pumpkins were unloaded ashore, and 'Ralph' — the old Land Rover given to him by the artist Ralph Hotere — helped tow them up steep farm tracks. Back at his workshop they were drilled out, and finally dropped onto bristling reinforcement rods to form the 30-metre columns of the Gateway Sculpture at the top of Victoria Street East, Auckland.

Amidst that 1987 labour, Chris had been asked to do the Rainbow Warrior memorial at Matauri Bay. Another Maori blessing for the big columnar basalts that lay on the Matauri shore. Another low-tech retrieval. Another spartan effort and another strong and symbolic basalt landmark.

We sat at lunch, talking rocks. Maybe we were talking too much about geology, too much about rocks, someone said and Chris replied

— Not at all. It's good to meet other boulder-brains.

The table talk drifted finally towards the stolen Te Araroa plaque.

— Just some yobbos out pig-hunting or something, said Chris. Someone has got that plaque hanging in their shed.

— I can get the plaque recast, I said.

— You've got all the words? said Chris. Great.

I had the words on a smartphone. I had a portable Bose speaker with a Bluetooth connection to the phone and it put out a full, rich sound. I'd just written the song 'Blue Trail — Te Araroa'. I set it onto the table and set it going.

A humble beginning. The sound of a lap steel guitar unwinding, as someone would later write, like fence wire along a boundary line. The bass doing its uphill and down. The footfall of the drum. The singer —

NORTHLAND

I could be happy
In blue and fortunate weather
Roaming the country
Between you and the sun

That was the start, and those were the words of Te Araroa's founding plaque. The first verse closed and the song swung into a Maori chant that picked out the placenames of Te Araroa's Northland route — Reinga, Ahipara, Orowhana, Mangamuka, Kerikeri, Waitangi, Mangawhai, Pakiri...

— Hey, it's a good song, said Chris at the end. And I love that waiata e hoa!

— The plaque should go back, and at some time the cairn has to go back, I said.

That question hung in the air, and we waited. Out to the right, just beyond the meadow, we could hear the river pouring over the Wharepuke Falls. Chris began to talk, and he wasn't talking to the table but to himself.

— It's still there, the cairn. It's a possibility, but I don't know. It could all fall to pieces. I'd have to have a look at it, but I think you'd snip whatever is still attached to the ground. There's only one helical cable, it's multistrand, and if you can get the cairn up in the air... Maybe you'd put wood through it before we start to bring it up to the vertical then try to open the bottom out while it's in the air and then drop it into a bed of wet concrete.

A sling attached to either end. The cairn like a pig on a spit, lifted, turned, repaired. The wet and waiting basement. I was delighted. I got ahead of myself.

— What do you reckon the cost is?

Chris registered the question and parked it.

— Well, that's always a difficult one.

Without pause he returned to the analysis of a reconstruction.

— And what I'd dearly like to do then would be to fill the whole thing with some sort of a medium filler, peat or something like that, and then once again the cairn becomes what I wanted it to become. Ferns and stuff growing out. I think for $5000 we could be on our way. We'll get it done as best we can.

— That's a wonderful price, I said. I'm sure we could get that money.

— Let's give it a go.

We went back inside, and piece by piece the family passed its rock collection in front of Mike. He examined a hard, flint-like rock and suggested silcrete. He examined an arrowhead Chris had chipped out of Kaeo obsidian with the same conchoidal scallops that marked Clovis Point arrowheads, turning it in his hands, impressed. He examined the rock with a trapped rainbow inside and declared a boulder opal matrix. He held a pale wriggled stone and pronounced a coprolite, then a fossil Anne-Marie passed across, and identified a gastropod. Chris tipped a jar full of shiny coloured pebbles clinking into a pile on the table.

— Chalcedony, said Mike. Sometimes you'll see a little pocket of stones in the peat, and so you think, 'How do you get highly polished rounded stones sitting in a pocket in what was a raised peat mire?'

— Moa, said Chris. Gizzard stones, all from Mitimiti. They were in one bunch on a very hard pan. I thought as a retirement project I could try to work out exactly where that moa walked to pick all these things up. Where did it pick this red one up from, and where did it pick the white one up?

— We think moa were our first rock hounds, said Mike. They picked out stones they were attracted to.

— Gemstone hounds! said Chris. If they were going to swallow stones, they may as well be attractive ones.

As we were leaving, Chris invited us into his shed. It was huge, with a wide roller door, a concrete floor set with stone cutting and drilling equipment, and a truck with a Hiab. Laid out in an oval pattern on the floor, or in heaps, drilled out and ready for weaving, were the stones for two cloaks.

Ralph Hotere died in February 2013, and Chris joined the tangi at Mitimiti, and listened to the debate. Some of the Mitimiti elders believed Hotere should be buried with his mother, a church person who'd been alone in Mitimiti's Anglican churchyard for 40 years. Others spoke for a burial in the Mitimiti urupa. Ralph should be with his tupuna. That impasse lasted through the first night, for and against, and sometime after midnight on the second night, one of Ralph's brothers got to his feet and lifted the debate up and over the deadlock. His mother's grave should be shrouded with a Chris Booth cloak, a kakahu with a Hotere motif, and up at the urupa, Ralph should have one too.

Chris and the Hotere whanau collected 4000 basalt cobbles from the Mitimiti shoreline. Taking shape now in the shed was a cloak with a design of concentric circles to warm the grave of the mother. A larger oval cloak for Ralph Hotere was still in planning. Probably it would centre around a thin white cross on a black ground.

— Not a Christian cross, said Chris, but his cross.

— The Maori solution meant a lot of extra work for you, I said.

— I owe it to Ralph, said Chris. I loved the man.

— Kerikeri is prosperous, said Mike as we drove south. It all comes out of the basalt soil. The money comes out of the basalt, but it comes out indirectly. You can see why the Anglicans liked this bit. The fertility. A nice little valley with ready access to the Bay of Islands. The Anglicans got here first and they got the good bit — the Catholics went and made their missions throughout the Hokianga Harbour where the soils are bloody awful and the land is that steep Cretaceous hill country.

Basalt had spilled over Kerikeri from 11 vents that erupted

mainly in the Pliocene and Pleistocene, five million years ago or less. From the Stone Store on up to its orchards and horticulture, Kerikeri was founded on basalt, and everyone there, in ways both mild or passionate, was beholden to it.

We drove to Waitangi, up past the Treaty Grounds to the Mt Bledisloe lookout, and looked out across the Bay of Islands' filigree coast. Waitangi too was part of the Kerikeri volcanic field, and lava rode out into the sea here in black shoals, but from the lookout the landscape opened out and a wider geology enveloped the local volcanic field.

— Most of those islands are basement greywacke, said Mike, and the greywacke is New Zealand's spine. It runs from here down to the Hunua Ranges at Auckland, Waiheke, down to Taupo to the edge of the Kaimanawas into the Kawekas, the Ruahines and the Tararuas. We're standing on that same spine — the axial ranges — and we're looking down there onto flatlands from the much younger basalt volcanoes.

— The young intraplate back-arc volcanoes erupted after all the allochthon excitement was over. Everything had stopped and then every now and again a basalt volcano would erupt, and they're not big basalts, they're not like Hawaii. But remember basalt flows like a river when it's hot and it might go ten or twelve kilometres.

— The Treaty Grounds down there, Mike pointed. They're on the lava flow of one of those basalt volcanoes.

On 6 February 1840, Lieutenant Hobson from HMS *Herald* is seated at a table and surrounded by Maori chiefs. The Treaty of Waitangi is unfurled and awaiting its signatures, and all of this grand theatre that is the founding of the New Zealand nation takes place on a wide and gently sloping basalt stage.

AUCKLAND
The volcanic field

Mid-February 1840. HMS *Herald* sails into the Waitemata to select a site for New Zealand's capital city. The ship anchors in the harbour, and as Lieutenant Governor William Hobson, Captain Joseph Nias, and the newly appointed Surveyor General of New Zealand, Felton Mathew, take the cutter out and about on the harbour, Mathew notes in his journal — 'There are several very singular hills rising boldly from the surrounding land . . .'

Three months later, the location of Auckland is still not decided, and Mathew explores the isthmus on the ground. By now, Mathew has recognised that the singular hills are part of a dense volcanic field, but he sees no threat. His wife Sarah is now newly arrived from Sydney, and the volcanic field is imbued only with a rosy glow. The tidally flooded crater at Panmure becomes, in his journal, a 'romantic little basin' and Maungarei (Mt Wellington) behind it a 'lofty volcanic mountain'.

It is Sarah who sees a starker landscape, noting in her journal the island of Rangitoto — 'surrounded by masses of black stone, resembling the ashes from some huge furnace . . .'

Later, in December of 1840, and January of 1841, as Felton Mathew climbs volcano after volcano to take Auckland's bearings, the power of the Auckland Volcanic Field begins to show in his sketches.

He names 'the Kings' — later called Three Kings — and as he lays out the shape of the new city he selects as his government precinct the high ground above Queen Street, part volcanic scoria cone, part ridge. The Governor's House is set up on the ridge, so too the church, the court, and the military barracks, and Auckland becomes the only capital city in the world to have its administrative heart within 200 metres of an explosion crater. Down below, Queen Street cuts through the crater's yellow outer rim.

In 1859 the geologist Dr Ferdinand von Hochstetter, visiting on an Austrian scientific expedition, is persuaded to stay on, and undertakes the first substantial survey of the volcanic field. Hochstetter is young and talented, and thrilled to find volcanoes around Auckland 'of an almost theoretical simplicity and perfect clarity'. His famous map of the isthmus shows an Auckland landscape pocked by craters, blanketed by thick tuff and overridden by lava. Almost half the isthmus has come under Vulcan's hammer.

In his later publication *Geologie von Neu-Seeland* Hochstetter remarks how often the basement rocks under Auckland 'have been broken through and pierced by the deep-seated volcanic power', and comments —

'The questions how long volcanic activity lasted on the isthmus and whether it could return once more naturally cannot be answered . . .'

In 2004, a geophysics team mapping the thickness of the crust under Northland and Auckland discovered the basalt melt under Auckland by chance.

The team, led by Nick Horspool, set out to measure upper North Island crustal thicknesses from Cape Reinga down to Tokoroa. It was the kind of project where geophysics excels — using instrumental tools over a wide field then analysing the data with computer algorithms. The team measured incoming waves from earthquakes as far afield as Papua New Guinea and Chile. The earthquakes were big — magnitude 6 and above — and the distance meant the waves went deep, before a small fraction of the rebounding energy came straight up through the substrates of Northland, Auckland and Waikato. That ascent was automatically recorded by six seismograph stations spread out along the upper North Island, and the geophysicists' job was to measure the velocity differentials, and translate those into depth and temperature for the upper mantle, the moho transition zone, and above that the crust.

The maths was complex, but the physics of it was simple. Earthquake waves travel at high speed through cold rock, and at reduced speeds through hot rock, in particular through basalt melt.

The study was broad-scale and not intended to show any great detail, yet when the observations had all been done and velocity gradients plotted onto a coloured graph, a blue-green stranger showed up in the mantle. I rang Horspool to check how his team had received the news.

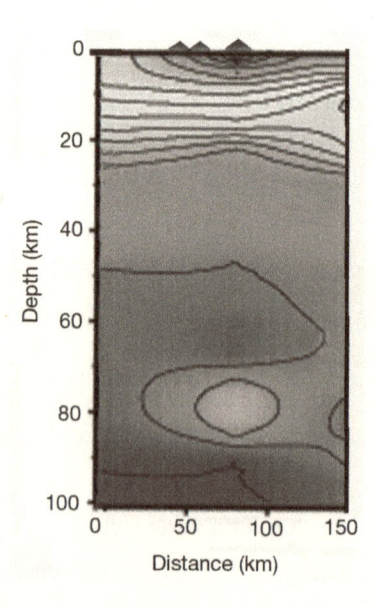

— We weren't looking for that, he replied, and when it did appear, we thought it might be an artefact of the modelling and tried to get rid of it.

The team checked its data and reran the programs. There was no error. At a depth of between 70 and 90 kilometres, an anomaly in seismic wave speeds caused by both temperature variation and melt content was slowing the earthquake waves as much as 20 per cent. The hot spot was almost 50 kilometres wide. Horspool says the 11 measured wave ascents through the field were insufficient to give it any exact three-dimensional shape, but agreed the 53 volcanic vents on Tamaki Makaurau's surface were a good clue. They traced a roughly oval shape, making it a fair assumption that the source melt below was similarly oval — the shape perhaps of a plutonic rugby ball, 50 kilometres long and 20 kilometres across. Geophysicists estimate only 2 or 3 per cent of that hot spot is actually partial basalt melt — the laces, perhaps, on that rugby ball — but if it unwinds, and the long thin string of it starts to rise, there's enough magma in it to give Auckland a huge surprise.

I was back in the city after the trip north with Mike Isaac.

Auckland seemed hot and happy and there were artists that summer willing to begin a conversation with the city's geology. At the Sculpture on the Shore exhibition on Takapuna Head, sound artist Charlotte Parallel had fitted out a World War II pillbox with a work she billed 'an auditory experience of geological systems'. The Geonet seismic and acoustic and pressure sensors around Auckland picked up every major urban vibration — the jackhammers, the earth-movers, the traffic, the westerlies, the waves that beat on the city's shoreline, maybe even the vast cheers that arose sometimes from the stadia, and certainly even the tiniest of the earthquakes. Parallel had converted the data from Geonet's 11 listening and motion stations to a sound file. White noise hissed ceaselessly in the pillbox, and if there was anything to take from the art it was a city that sat with its ears pricked 24/7 listening within its cultural noise for the deeper notes of ascending magma. Parallel's work was called Monogenetic Field, which was the geologists' formal description of the AVF — the name for any field of single-shot volcanoes whose plumbing hardens over time, forcing each new magma pulse to find a new way to the surface. Monogenetic volcanoes can come up anywhere within the field, and in that eventuality, given Auckland's wall-to-wall real estate, the danger to life and property would be extreme. And so you listened, and hopefully could accurately triangulate that approaching rumble, and maybe even have time to evacuate a five-kilometre circle around the new eruption point. Meantime all of the cultural noise of the city and its substrate hissed and crackled, and the kids raced in and out of the pillbox, and no one in Auckland seemed worried about any of it.

Also, I checked back in with the Cabin Fever Club, a loose group of locals who work from home, alone, and meet weekly for an hour at a Devonport café. The club is entirely informal, there's no need to say you're coming and, as a result, just occasionally there's the so-called 'Quorum of One', who nurses his thoughts and his coffee alone for an hour and returns home. Most weeks, though, six or seven show up, turn over the latest news and discuss their various projects.

— What have you been up to this week, Geoff?

I tell them about the rugby ball under Auckland. That we should hope that no one puts in a high up and under any time soon. I always try to keep it brief. I know I can hold their interest, but at the same time the geological bore is a comic trope. I'd watched, as had a few of them also, the recent British comedy duo Steve Coogan and Rob Brydon playing geology for laughs in their BBC documentary *The Trip*. Brydon just wants to look at the

Yorkshire landscape, to quote Wordsworth, to hit on the Yorkshire sublime with a 'Wow', and it's Coogan who interrupts the reverie time and again with his tiresome litany of rock formation, age, time, and place.

One CFC member, its founder no less, is alert to every comic trope. Roger Hall — playwright. I watch him gauge the various CFC members for their comic potential and for a possible disguised use in some new play: the geek, the poet manqué, the disgruntled novelist, the out-of-work actor. None of us is quite those things, but if the character was pushed . . . I can see the gleam in his eye as my own caricature emerges, and he tests the possibilities. I talk rocks, in respect of geology I'm the Coogan of the group, and there's a running joke whereby, after a minute or two, Roger will reach out with two hands as if to clamp, uplift and transport me.

— The Geology sub-committee will continue to meet . . . over there.

Swinging the clamp around to deliver me to an empty table, distant as a desert isle, where, the suggestion is, the Geology sub-committee of one can talk its head off.

— What do you make of this? Warwick Freeman took a big hunk of grey-green rock out of its plastic wrapping and laid it on the table in front of a full meeting of the CFC.

Warwick was the newest member of the group. Roger had met him by chance on the Devonport ferry, started talking, found him interesting, and confirmed he worked alone. The group needed to okay any new member, Roger gave out Warwick's website address, and I looked it up.

He was a jeweller. He'd once been part of the Fingers co-operative, a loose collective whose members worked independently but under the sway of a single idea — to base its work on New Zealand design and materials. And then there was something amongst his list of work that made me sit up straight: *16 Ball Rings, North Cape to Bluff*. The rings were oxidised silver, or gold, each one set with a small stone globe as neat and smooth as if it'd been separated from its parent rock by a melon scoop. Each globe came from a specific point down the length of the country, samples of the wide variety of New Zealand's geology.

Now here he was, with a large sea-smoothed rock picked up on the Devonport shoreline. I took hold of the rock and weighed it in my hands, heavy, close-grained and very hard.

— It's not greywacke.

— I think it could be argillite.

I turned it over. The rock had a green tinge and a couple of limpets clinging to the underside. I agreed with him.

— Yes, argillite.

— What's the difference, said someone, with the greywacke and the other one?

— Greywacke is a sandstone, I said. It's hardened into a dense rock at depth. Argillite is a mudstone, also buried at depth, so maybe just as hard.

Warwick suddenly whipped out a geological hammer and banged the rock. A big chip split off. A puff of rock dust drifted up.

— Yes, argillite I'd say. It's tight, and you can see it breaks in these conchoidal fractures.

In that happy moment, by that single stroke, the Geology sub-committee of one became two and I wasn't even sure that I retained the status of sub-committee chair.

A few weeks later I went up to Warwick's workshop. It was a hack space to dream on, standing separate in a big yard just beyond the vegetable patch, with wide windows to let in the light, and the imposing symmetry of Rangitoto standing behind, sunk, as the poet once wrote, in its stone composure.

The argillite was on the bench, already sliced to reveal its dark interior, along with chalcedonies, and jaspers, and lapis lazuli in various stages of reduction and reassembly. There were presses and clamps, a circular saw without teeth beyond the minuscule grinding edge of a blade that was edged in diamonds, and a whole separate annex for the diamond grinders, and the diamond hole saws that cut a neat circle through rock until you pushed them just that fraction too long, and *phwoosh!* the diamond edge turned to gas. It was a great workshop, and also one redolent of a bloke's shed, the grinders driven by electric motors poozled from old fridges at the Devonport dump.

One box contained chunks of twisted basalt. Warwick was experimenting with the rock as a natural shape that might make a useful door handle, or drawer pull. They'd come from volcanic bombs, the bombs themselves long gone, but broken bits of them still lay scattered on the black reefs around Devonport.

There was the aerated basalt of the Auckland Volcanic Field cut in thin section, not for any geological analysis of its mineralogy, but for the odd beauty of its interior.

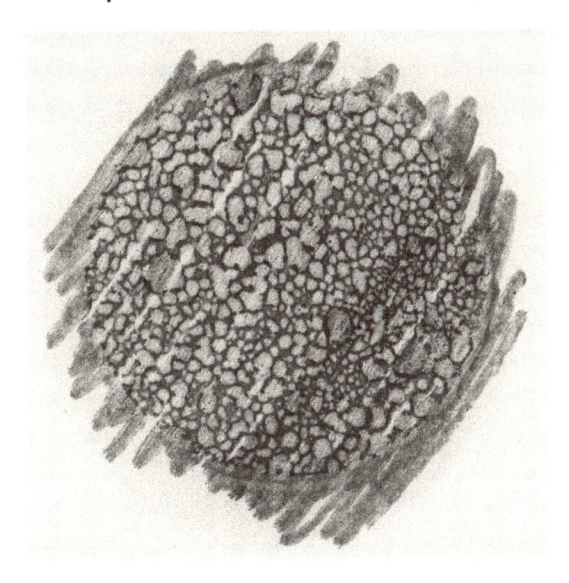

— So. You study geology?

— Not the science, but the social thing. How people respond to the rock, and how they use the rock. That's my interest.

I said I thought New Zealanders in general responded to geology. Maybe it was just the force of the tectonics, Christchurch's terrible quakes, the recent interest in the sunken continent Zealandia. Whatever, they had an interest, and the interest seemed genuine.

More than that, suggested Warwick. New Zealand's history was so brief, each of its separate populations had invoked a powerful story for the land that stood at their backs. Not just the land that existed now, which was wild and could be momentarily fierce, but the primal land of the past whose dynamic movement made it wilder yet. These things became a natural part of New Zealanders' conversation, for the peopled part of their history went back only 750 years. Maori had used their gods to journey back beyond that, and Pakeha did the same thing with geology. Simple.

The tide was unusually low next day, and we moved across the reef at the base of North Head, looking for the twisting spindles of basalt, shaped during flight or from extrusion under the sea.

— Occasionally, said Warwick, you see something that sets the pulse racing.

He dislodged a smooth rock out of the mud, turned it over, cast it aside.

— If you got lucky, it'd be an adze tossed out of a waka coming ashore here.

— You mean dropped.

— Maybe just dispensed with. I know the current view is to regard anything made by pre-European Maori as taonga, but I still see objects like the stone adzes as part of an industry. Just a tool. If you were bringing down a large tree, you'd break adzes in the process and I think it's like, 'Oh shit — whip up to the quarry and see if they've finished some more — this tree is going to take at least six.' That's the thing you'll never know, the line between how they lived with their industrial culture. We put it out in our rubbish bags. Was it any different then?

When we finished collecting, Warwick had over 30 pieces of basalt that held some semblance of its plastic form as cooled in air or cooled by water. He had his eye in from long experience, and I was the novice and contributed just five. Back at the workshop they'd be stripped of their clinging shell in a hydrochloric acid bath, then oiled, cross-cut to fit one end of them flush with a drawer or door, and the other end left to protrude as a handle.

Warwick's interest lay in how stone got used socially, and I suggested we take a walk across Auckland to see what else we might spot on Te Araroa's route across the city.

The journey cross-harbour on the Devonport ferry began in pandemonium. The Kaitataki house team from Alfriston College wore yellow T-shirts with a lion on the front, the legend *Work Hard — Play Hard*, the big black letters *KTK*, and they'd just gone OTT after spotting Devonport's Ella Maria Lani Yelich-O'Connor, otherwise known as the new mistress of pop, Lorde. The group crescendoed on one side of the boat when someone spotted her, then crescendoed again on the other side when she greeted them. They had certain rights that included crowding around, jumping, clapping hands, and tossing their long hair to match the waterfall of Lorde's own locks. It struck me that 'fans' was the wrong word for them. They were her clients, and the job they ceded to her was to make music and craft the image. No one was about to mob Warwick Freeman, but at the same time he had his clients — white, urban townspeople — who'd ceded to him a similar licence, to investigate New Zealand rock and to suggest adornments that might appeal to them as white natives.

We left pop culture behind on the city-side wharf and at Emily Place stooped instead over old basalt kerbstones knapped by forgotten Mt Eden prisoners. On Princes Street we looked up to the university clock tower's Mt Somers limestone gothic, then laid hands on the last remnant of the 1840s Britomart Barracks fort preserved within the university campus, a wall with rifle loopholes, made from basalt with a grout of pipi shells. On Grafton Road we passed the Owen Glenn Business School, and Warwick ran his hand over exterior panels of stone chip. Bluestone, he thought, probably an Auckland basalt, but the big panels sat there without the mortised joins of traditional stonemasonry, locked onto an unseen steel frame, and separated, where you might expect mortar, by the hard, squeezed rubber of modernism. It was a far cry from the solid bluestone walls of Auckland's Kinder House and the Anglican churches built by the master masons of the nineteenth century.

— There's a hierarchy of materials in architecture, said Warwick, and probably in the late nineteenth century that hierarchy said bluestone was worthy. Not flash maybe, but a worthy material. The churches, and I'd guess any building of a certain quality, were worthy in that way.

Here, in the twenty-first century, the bluestone had gone thin and clever, a shallow sprinkle on the enormous exterior panels of the business school, but nonetheless a tip of the hat to the city's hard volcanic stone.

We were headed up the Auckland Domain under a canopy of pohutukawa, trying to recall more of Auckland's memorable stone, when suddenly Warwick spoke.

— I was trying to remember where I've seen one of the great scoria decorative walls, and of course it's the Central Police Station.

— Hmmnnn. And something about scoria suits the police station.

— That's right. 'Tell us, or it's the scoria wall for you, mate!'

It started with rock chips on the Kina Peninsula in Tasman Bay. His parents had a holiday bach there and, as a child in the 1960s, Warwick found the argillite chips.

He'd stumbled on something. He didn't know what it was, except that it focused a young wish for some single stroke of discovery, some three-column headline in the *Nelson Evening Mail* — 'Schoolboy uncovers astounding relic of ancient race' — and although he did find a few pre-form adzes, the real discovery was no single find, but an assembly of thoughtful encounters over years. No single adze, but evidence of a production line that encompassed thousands. He began to see the scale of the neolithic industry around Tasman Bay, and it wasn't an ancient race. It seemed like it had happened yesterday.

Countrywide, argillite was not a useful neolithic rock, but the argillite that surfaced in monoliths around Nelson had been forced deep by the vast tectonics of the Nelson mineral belt, hardened with minerals, then delivered back to the light. It was fine-grained, homogeneous, its textures so even that any blow struck against it was a cone of force that raced evenly through the rock and broke pieces from the body of it in the shape of a fan — the conchoidal chip. Such easy control of the chip allowed you to rough out the shape of a tool quickly, and the argillite was so hard you could grind an edge.

Maori called it pakohe. Warwick found quarries up the Maitai Valley behind Nelson still blackened at their base by fires built to bust useable chunks off the argillite monoliths. The quarry workers had trapped a water source at the Rush Pool, and the ground there was littered with chips. They'd shaped up blanks, the first preparation for transport, and packed them out 16 kilometres downriver, and another 20 kilometres by waka across to the finishing sites on the peninsula. The argillite from D'Urville Island 65 kilometres away came in on a different route, by waka straight across the bay. Warwick explored all the quarries. They extended, sometimes, whole hectares.

Aotearoa had one of the last neolithic cultures. In Europe the culture had vanished thousands of years before — in New Zealand it tapered out less than 200 years ago, much of it still talked about and remembered. Maori knew their stone. They also knew when a more useful material appeared, and post-contact they dropped the stone and took up steel. They called those who brought them the new tools Pakeha, and for a time the label was asymmetric, and then it was not.

Pakeha. Their name for him, his name for himself. Warwick felt a respect for the Maori quarrymen and finishers that was centred on craft. He felt no particular divide and he called them by a name neutral to all of them: 'makers'.

— When people say 'stone age culture', it's pejorative almost. It's used to describe a lack of rather than a respect for, but it's a technology that I can grasp and understand and appreciate.

In the 1980s, a group of Pakeha stone carvers set out to bring pakohe back. John Edgar from Auckland's Waitakere Ranges had explored the Nelson source quarries, and in 1984 he mounted an exhibition at the Dowse Gallery in Lower Hutt that featured the rock. Warwick Freeman was amongst the exhibitors. A Pakeha maker. He put in necklaces made from argillite flakes.

Know who you are by where you are: Pakeha culture took its references and confidences from the Kiwi vernacular in literature, Kiwi visions of red pohutukawa blossom against a blue sky in the plastic arts, but the self-aware knitting in of the rock came later. Nineteenth-century Maori makers had ascended out of the stone age with the tools supplied by European contact, and then New Zealand stone would wait into the 1980s before European makers would take the tools and descend.

— The whole of Pakeha culture is much more rudimentary than, say, something out of England, said Warwick, and when you're taking your identity from something as rudimentary as a rock, you've really stripped it back. Okay, it lacks a certain sophistication, yet at the same time people like Michael King made us appreciate that it might sound rudimentary but in it is also the complexity of who we are.

In gratitude and grief for its war dead, the Auckland region had pooled its money and imported for its War Memorial Museum the same gracious limestone as St Paul's Cathedral in London, English Portland Stone. The style of the building, with its huge Doric columns, its pediments, the spacious atrium, is classically Greek, as channelled by eighteenth- and nineteenth-century Western architects. There's hardly an original bone in its large rectangular body, but it is huge and solemn and does its job well, of remembrance and respect for the great intellectual traditions of the West. We went in to look at the local stone age culture.

— That's a typical tree-felling adze, shaped completely by flaking.

Warwick was looking at a half-metre piece of tapering rock. It was behind glass, and you couldn't touch, but just a look told you it was big and very heavy.

— If you look at that from all angles, you'll find the lines are pretty controlled. From what I understand they were hafted onto a long piece of timber and that was secured with a rope that went up fifteen or twenty metres, and acted like a pendulum so you could swing it straight in, and chip away.

We turned to the adzes, which had been chipped in the rough, then polished, and Warwick was pointing through the glass at an adze made of D'Urville argillite, a pale grey flecked with black, a top-quality stone.

— They're risking a much thinner section with this than they are with basalt. They're using a much finer working material, so they have a lot of control.

Then the shining pounamu adzes.

— If you try and chip those, they won't crack on any predetermined line. They'll crack on a flaw. So they require an entirely different technology, a sandstone saw, back and forward in a groove, with water and sand.

We moved to the hei tiki display, a dozen green and grimacing charms, heads aslant, and Warwick looked for, but didn't find, the museum's nascent tiki that protrudes like a bas-relief from its parent adze. The tiki is only half done and, for whatever reason, its maker never completed the work. For Warwick, the prefigured tiki on a greenstone adze marked the long moment of Maori transition away from stone tools. European traders wanted hei tiki, not stone axes, and the Maori carvers responded by turning the redundant adzes into a more figurative, tradeable item.

— The beginning of the souvenir trade, he said. So many of the hei tiki have that adze shape with the head aslant, and it makes practical sense, too. I think it could be linked to economy. You need to remove a lot less material with the head on its side like that.

The Pakeha maker was assessing the Maori makers. Call it jewellery or call it adornment, but Maori had also shaped their precious greenstone into the elegant kapeu, the ear pendants that were closer to his own use of stone. We looked at those, and before we left, Warwick took me to a case where his own jewellery was on display. A scoria lei made in 1989, the hard edges knocked off from being tumbled in the tide, dozens of small knobs of the stuff, spindled with silver. Local rock, a welcome from Auckland, a necklace for Auckland. A Pakeha response to Tamaki Makaurau.

We left the museum for the Domain playing fields. The old volcanic craters of Pukekawa were now converted from swampland and rolled flat as sports fields, volcanoes groomed as local stadia, enclosed by their tuff rings, now nicely grassed as an encompassing vantage for spectators.

On through the oaks, and down to the small square pediment of a Gallipoli memorial. Cobbles from Gallipoli were stacked side by side with local stone. Mustafa Ataturk, the president of Turkey, had offered a reconciliation in the 1930s for the bloody British attempt to capture the Dardanelles during World War I. *Our Mehmets and your Johnnies,* Ataturk had said, *lying side by side.* Our rocks, your rocks, rocks used to plumb death itself, the single unity of war, but were the ratios right, or like so many memorials, adjusted to the severity of your own losses? The Turkish defenders, with inferior weaponry but a deathless determination, lost twice as many men. Around 86,000 Mehmets were killed in defence of their country, against 44,000 of the invading Johnnies. New Zealand troops made up 2779 of that Allied number.

Mountain Road was lined with basalt walls — old ones, and new ones built by Tongan work gangs. We walked past Auckland Grammar School with its distinctive stucco of broken and protruding stone, and we were looking for a rumoured piece of original Auckland lava field. As we came abreast of Government House and turned down Withiel Drive, Warwick was first to spot this small wild park through the trees.

— That looks like unreformed basalt rumbling down.

The entrance was unsigned, but we passed into a shadowy interior where huge black blocks lay tumbled together. For a moment we were back with the first settlers in a primal landscape.

— It's original, I said. It's impressive, but there's nothing much here for a jeweller is there?

— You never know, said Warwick. Where do your ideas come from? There is no answer to that. You get them from being alive, so I never second-guess any experience.

The roots of large trees writhed over the black piles, and delved down between them. Where do ideas come from? Back at the turn of the millennium, New Zealand proclaimed itself the first country to see the sun, and an Austrian collector wrote to Warwick. He was mounting a themed show, 'Jewellery at the Turn of the Millennium', and wanted a New Zealand contribution.

'How is it possible to see time?' the collector wrote, setting the Kiwi jeweller a challenge. 'Maybe New Zealanders have a better vision.'

Warwick thought about that. The whole millennial razzmatazz seemed entirely arbitrary to him, and for a while he was adrift. He needed something tangible, and he couldn't find it in the claims of which country was most immediately west of the international dateline, nor whether a Chatham Island or a Hikurangi mainland mountain was best to catch the first rays of the rising sun. But the geology was solid. He thought himself back 2000 years, which was, after all, what it was all about, and found, on the last of the world's big islands to be colonised by humanity, that there was no one there yet. He could see the flickering of the day and night rhythm, light and dark on a nameless place, but really, without humanity, there was no time. Detonations, yes, they were real. The shining lake that would report to the future as Taupo Moana was within a geological tick or two of an eruption that would overshadow the world, and the pressures were already sending volcanic glass hissing along the margins of smaller vents, and feeding the thermal upwellings that brought precious metals to the surface.

And if you had to introduce humanity into the fraudulent moment of the millennium, then throw a few lines from Janet Frame's poem into the mix —

> *To undeceive the sight a detached instrument like a mirror is necessary.*
> *The human senses never speak the truth if they can get away with an easy lie.*

The Pakeha jeweller produced for the Austrian collector a black mirror, a disk of polished Taupo obsidian, 2000 years old, and banded with gold. A vanity mirror. Look into that, toss your millennial curls, or practise your happy face, and all that stared back was the mineral world, dark and timeless, implacable.

A Maori docent wearing Auckland Council insignia stood on the Maungawhau summit. The summit is part of Te Araroa's route across Auckland, and he'd just seen off a walker going the other way. He turned to me.

— Do you have any questions about volcanoes at all?
— Yep. When's the next one?
— Okay. You'll have to ask some greater power . . .
— Right. I'll tap into Ruaumoko.
— Well no, said the docent. He's the earthquake god when he kicks, and he can also be the volcano god, but in this region the god's name is Mataoho.

He gestured towards the large green crater countersunk just below our summit viewpoint.

—That is Te ipu kai a Mataoho. His food bowl.

We stood on the summit and looked out on an isthmus that had boiled like a pot of old porridge for something over 200,000 years. The big green lumps of old volcanoes stood up all around us, but the volcanic island that stood at the harbour mouth was not only an obviously different shape, larger and more symmetric, but also a different colour — battleship grey. A geologist who'd seemed exasperated by its refusal to conform to the patterns of the AVF had once described it to me as 'weird'. It was the only bad word I'd ever heard against Auckland's greatest icon. The city loved it, but it was also the only one that could, if the bulk and power of it ever recurred, break the city's heart.

'Rangitoto,' wrote Dr Ferdinand von Hochstetter in his 1864 *Geologie von Neu-Seeland*, 'is for Auckland Harbour what Vesuvius is for the Bay of Naples, the "symbol" of Auckland.' Hochstetter described Rangitoto as small compared with the world's more renowned shield volcanoes, but nevertheless 'a true model of a volcanic mountain cone'. In comparison with the pocket volcanoes that dot the isthmus, though, it was huge, disgorging two cubic kilometres of lava and spreading over some 30 square kilometres. It drew magma not just from the mantle, but also from a shallow crustal melt. Unlike the rest of the AVF, it erupted twice, with a decades-long gap in the 550–600-year period, and there was a suggestion, not yet proven, that it had also erupted for the first time 1000 years before that. You had to hope the volcano gods would be kind and, when the time came, offer up another North Head or Mount Roskill, and not a Rangitoto, for in the wrong place it would close down most of the isthmus.

The docent was sweeping an arm east to west. The god's long-ago Battle of the Sunbeams had caused the eruptive mounds along that axis. Then he swung his arm back to the north–south axis where Te Mataoho's bodyparts had ruptured the surface.

— Mangere Mountain out there, that's Te Pane o Mataoho, the forehead of him sticking out. And you see the industrial area out there? That's where his nose pops out. Te Ihumatao o Mataoho. That's the cold nose of Mataoho. It's quite windy out there — quite chilly.

The day was getting on, the sun's shadow creeping across the crater below us. If we walked on we were going to miss our planned low-tide walk around the Manukau shoreline, and we decided to short-cut. Warwick got on his cell phone to order a cab, and Despatch asked where he wanted to go.

— Te Ihumatao. Out by the airport. The cold nose.

— True? said the docent. Well, while you're out there, get some avocados from the free orchard.

The avocado orchard had no low-hanging fruit, and we threw thick sticks up towards the tops, knocking the upper branches until some of the high fruit dropped. They were small and hard, but they were free. We walked on where Maori had once laid down huge gardens, radiating warmth back onto the kumara, the yams and the bottle gourds by heaping scoria alongside into useful alignments. We walked south through the fields, down towards the coast, and Maungataketake was a kilometre away across the fields.

Maungataketake was a model for any one of the small volcanoes within the AVF. A nicely raised tuff ring in a circle around the blast crater, then, rising within the kilometre-wide reach of that wide flat crater, a scoria cone. In the days or perhaps weeks of its fire-fountaining, Maungataketake had heaped up a 75-metre central cone. The long thread of magma, up from the depths through that central throat, held its CO_2, fluorine, sulphur, and H_2O volatiles in liquid compression, but as it hit the vastly decreased pressure of the surface it degassed with a roar, and the escaping volatiles riddled the rock a thousand times through. A basaltic froth that rose high and fell back to earth as scoria.

Apropos of nothing at all, Warwick said

— Try spelling scoria backwards and see what you get.

I had a go and didn't get there.

— Air rocks, he said.

— You have to use the 'r' twice, I said. There's no 'k'.

— Airocs, he said, spelling it out. Okay, spelling isn't my greatest strength. But scoria should be the regional rock, because Auckland is built on it. And better the city of volcanoes than the city of sails. It's more connected. Not everyone has a yacht.

Maungataketake built its scoria cone, and then, when the pressure subsided, oozed lava, a classic volcanic progression, and if you wanted to add culture, it came complete with Maori fortification. Hochstetter had responded to the mountain. His *Geologie* praised it as the most important of the southern cones, and Charles

Fleming, in 1959 the translator of *Geologie*, attached a footnote that the mountain 'is one of two or three cones still completely intact'. White's Aviation photographed it still whole in the 1960s, but as it came into view now, it was a carcass, flensed almost to ground level. You could see the dark red interior, stepped down in terraces where the dump trucks lumbered through. This one, and many before it, providing scoria for the city's boundary walls, foundations for its tennis courts, substrate for its railway lines, scoria for whatever drainage pit or channel you might want tricked out with a hard but permeable fill. Auckland was necklaced with it a thousand times around.

We walked to the coast across the rippling lava that spilled westward from Otuataua, climbed fences to reach the shoreline and stepped into the wind. Insofar as you can polish basalt, the edge of the Otuataua lava flow was polished by the blown sand of the westerlies. The faces of it gleamed dully, like lead.

We fossicked slowly on around, looking for remnant lava bomb pieces. We rounded a headland, and the Auckland International Airport runway came into view just two kilometres south, extending out into the flats. We rounded another headland and walked towards the low yellow cliffs.

— Here's an Indian god.

Ganesh was staring up at us from the mud. Multicultural Auckland, but by then we were near the end of Renton Road. Severed tree bases and the trunks and branches of an old forest stood in the cliff, destroyed by Maungataketake in its first eruptive blasts.

We laid hands on the cliff. It was layered with black pellets often no bigger than buckshot, grading up to fine sandstone, then another layer of pellets, grading up, then another. The city had stripped away Maungataketake's cone, but no one could take away the tuff it had flung outward, layer after layer in concentric aprons around its blast crater. Again and again in the first days of its first eruption, before the scoria cone, before the lava, Maungataketake had boomed like a bell. I turned to Warwick.

— Geologists love this volcano. They can't easily get at most of the Auckland Volcanic Field because it's covered with houses, so Maungataketake is the stand-in. They've done the ground radar to trace the runouts, they've dug the forest. They've put in a report.

The first magma hit the water table here, flashed it to steam, and disintegrated into fragments, taking with it the wide tonnes of macerated Waitemata sandstone. Those base surges rolled away from the crater 60 metres high. They rolled at around 230 kilometres an hour. Every cubic metre of the surging cloud was packed with 38 kilograms of broken basalt and sandstone. Close in, the basalt shrapnel severed the forest at its root, and one kilometre out from the crater was still felling trees half a metre wide. The blast exited at 1000°C and was still a scorching 500°C when it hit the outskirts. Such blasts, transposed to any modern suburb, might spare your house its complete destruction beyond that first kilometre, but any one of the blasts might overwhelm you anyway with heat, and by its concussion bowl you over even two kilometres distant. To complete the carnage, ballistics would rain down redhot and ignite fires, or come down cold and knock your brains out. That was Maungataketake, a far smaller volcano than the Mt Edens or the One Tree Hills, or the Three Kings that lay further north. If you wanted to know what happened there, then no one exactly knew, but you could do the maths.

A Boeing 747 hung in the air, climbing away, big and heavy, impossibly slow. Warwick was gazing at it in his purposeful way.

— It's such a big energy in an island, kind of bottom-of-the-world culture. It's the way out. Out to the Manukau Heads, and crank a right.

By then we were finding bright red sequinned scarves draped on rock. Yellow pennants and Sanskrit prayers blown into dead trees. The tidal wrack was littered with small terracotta bowls blackened by flame. With coconuts, still whole, their matting soaked and plastered with a dank decoration. Another god, headless, but one hand upraised, palm out. Another one, metal this time, encrusted with mineral salts, corroding.

We were both trying to construct some kind of geomantic explanation, beyond Maori, beyond Pakeha, to the new cultures that had come into Auckland and were finding their way in the landscape. The planes. The volcano. The tides. The bowls of lighted camphor taken by the tide out towards the distant stone gates of the Manukau, and if you went on through that gap, cleared Australia and the ocean beyond, you might descend upon the fabled coast of Malabar.

We rang a cab to come through from the airport, and when it arrived, the driver was Indian, a former policeman from Fiji. I asked him what was going on down on the beach.

— After the death of a person they take the family gods to the seaside where they do a small ceremony. They shave off their heads and all those things and then make the offerings to the person who died and to the gods they believe in. And then they put it into the sea. They believe the sea is connected to everywhere. So. That's the way it is.

EAST COAST
Subduction

The United States Navy polar research vessel *Eltanin* sailed into Auckland for a refit in December 1964, and the *New Zealand Herald* duly ran a story under the headline 'Picture Puzzle from Seabed'. It was a photo from the *Eltanin*'s undersea camera, taken 1000 kilometres west of Cape Horn, and showed an antenna lookalike, a thin upright stem with regular crosspieces. *Eltanin* scientists couldn't identify the object, but Auckland's busiest UFOlogist NAC pilot, Captain Bruce Cathie, came aboard the ship to check it out, and was sure he knew. A few years later, about the same time the mystery object was identified as a carnivorous sponge, Cathie published his best-selling book *Harmonic 33*, alleging the antenna was part of a worldwide power grid where UFOs refuelled.

You could argue that both Bruce Cathie and the *Herald* had missed the big story, which was *Eltanin*'s refit with a new Sparker seismic reflection unit and upgrades to the magnetometers. The ship was preparing a significant test for a revolutionary new theory that would come to be named plate tectonics.

In 1912, the German geologist Alfred Wegener had proposed that the continents showed a suggestive fit at the 200-metre isobath line. His evidences included the similarity of continental rock types and fossil assemblages either side of a mutual ocean, and evidences of polar wander. It was good enough to hang your hat on, but it didn't explain how the continents might plough through ocean bedrock.

Research relevant to the theory continued, including work done by Felix Vening Meinesz in Dutch submarines charting gravity anomalies in ocean trenches off Indonesia. The Princeton geologist Harold Hess worked occasionally with Vening Meinesz and joined the US Naval Reserve in the late 1930s, at least partially to enlist American submarines in the exploration of oceanic trenches, but then war overtook the world and Hess became skipper of the American troop transport USS *Cape Johnson*. American troop transports landed GIs in the shallow water of the Philippines war theatre, and were equipped with fathometers to signal water depths under the keel. The fathometer wasn't unusual, but the *Cape Johnson*'s skipper was. Whether under Japanese rocket fire in the Philippines, or on his voyages to and from his Hawaii base, he never turned his fathometer off.

Immediately postwar, oceanography took on Cold War urgency and military budgets rose to suit. Hess led a team correlating screeds of bathymetric data from other Navy ships. Whatever other results, the Navy wanted hiding places for its submarines. The military also wanted a seismograph network sensitive enough to detect Russian or Chinese nuclear tests, and military budgets helped salt the new Wood-Anderson seismometers around the world.

Over the next decade, oceanography made huge advances. Research ships mapped mid-ocean ridges, which ran like seams through every major ocean. Seismologists Charles Richter and Beno Gutenberg produced global maps showing quakes clustered along ocean trenches and mid-ocean ridges. Hess, by now a rear admiral in the Naval Reserve, was in a pivotal position. He had the rank, and he had the science. A cascade of specific undersea information passed across his desk. Sediment was thin close to the ocean ridges, but much thicker near the trenches. The rifts in the mid-ocean ridges were shedding heat from linear upwellings of molten rock. And then came the critical geophysical information. Scientists already knew the earth's magnetic field changed polarity on a million-year time scale, and that iron-rich igneous rocks captured that change. They'd already fitted the regular reversals — over 20 even within the last 5 million years — into the geological timescales of Pliocene, and Miocene, and beyond. Those had terrestrial uses, but now the magnetometers towed behind the research ships were charting extraordinarily regular stripes of normal and reversed polarity across the ocean floor. The younger stripes were close to the mid-ocean ridges, and the older stripes were further away. That seemed to suggest the ocean floor was spreading.

In 1960 Hess stitched it together and produced for the Office of Naval Research a paper that outlined, for the first time, the rudiments of what would become plate tectonics. His paper paid tribute to Wegener, who'd died in 1930, to the British geologist Arthur Holmes who'd kept the idea alive since, and to Vening Meinesz. Hess had enough evidence to lift his interpretation high above those early researchers, but even so he had to make large unproven assumptions, and the paper was playful around the fact. Hess called his paper an exercise in geopoetry.

He proposed that ocean plates were continuously generated at mid-ocean ridges, and continuously subducted at oceanic trenches, a cycle of creation and destruction that would completely renew the ocean floor every 300–400 million years.

He knitted in the flat-topped undersea volcanoes he'd discovered during his wartime voyages to and from Hawaii and that he'd named guyots. They'd been wave-planed across their summits as they began life poking above the sea at the mid-ocean ridges, he speculated, then protected from further erosion as they journeyed down to abyssal depths, and ended finally in what he called the 'jaw crusher' of the descending plate.

'The continents do not plow through oceanic crust impelled by unknown forces,' he wrote, sidelining the objection that had dogged Wegener's proposal, 'rather they ride passively on mantle material as it comes to the surface at the crest of a ridge and then moves laterally away.'

The paper was formally published in November 1962, and for the next two years was the most-read scientific paper in the world. It was speculative geopoetry, but it could be tested in a number of important ways. In 1963, the Cambridge geophysicists Frederick Vine and Drummond Matthews, and independently the Canadian geologist Lawrence Morley, recognised that if Hess was right and new sea floor was being continuously generated by eruptions along the mid-ocean ridges, then the magnetic striping either side of the ridges should present, in ideal conditions, a mirror image.

Among the illustrations in Hess's paper was a diagram of the East Pacific Rise, which snaked below New Zealand in stormy 50° and 60° south latitudes before turning north towards the equator and intersecting North America near the San Andreas Fault. Hess captioned it 'an oceanic ridge so young that it had not yet developed a median rift zone'. The East Pacific Rise was not, like the Mid Atlantic Rise, positioned squarely between two continents, but it was young, and might be less distorted than ocean ridges elsewhere. The Americans had a research ship in the area. The call went out to the *Eltanin* and the Americans flew in new kit to prepare the ship for its task.

After the Auckland refit, the *Eltanin* tested its new systems with a month-long cruise out from Auckland, went to Wellington, then in 1965 began the cruise that secured its place in history. The ship's 1000-kilometre run across four degrees of latitude from 54° to 50° south produced a near-perfect mirror image of magnetic stripes and depths either side of the East Pacific Rise. The theory of plate tectonics was still unnamed, but the geopoetry was becoming hard science. The *Eltanin* image would begin to turn a majority of geological sceptics into a majority of believers, and it became one of the most influential geophysical profiles ever published.

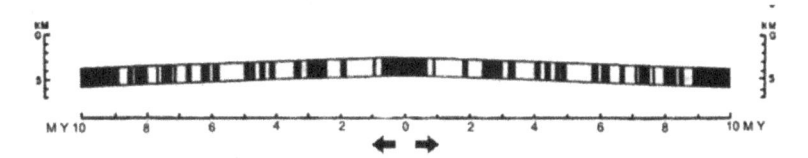

Big science was at work on the oceans of the world, and New Zealand was beginning its own offshore surveys around the East Coast. The *Eltanin* assisted the effort by design or by default. Two young employees from the New Zealand Oceanographic Institute, Keith Lewis and Jeremy Gibb, were regular visitors when the ship was in port at Wellington. New Zealand seabed samples taken from piston corers were often compressed and damaged during extraction from the steel pipes. The Americans used plastic sleeves that could be pulled from the coring pipe then sliced to reveal the sediments in nicely displayed half-rounds. The Americans also seemed to take for granted their Maxwell House instant coffee, but the Kiwis had never tasted anything like it, and it beat Bushells Coffee and Chicory Essence. The two New Zealand geologists took the Americans hunting in the Tararuas and Orongorongos, and the payoff was plastic liners for the piston corers and tins of Maxwell House coffee.

On a more formal level, the *Eltanin* seismologists had given New Zealand a small but interesting undersea profile of the Hikurangi Trench. Keith Lewis had studied that before he put to sea on 9 April 1968 in the old Fijian copra trader MV *Taranui*. It was the first day of a PhD thesis to probe more thoroughly a puzzling series of undersea ridges off the East Coast. The interisland ferry *Wahine* followed them out, and passed the *Taranui* in the night. On 10 April, Keith woke to storm-force winds. The *Taranui* was hove to off Kaikoura, and the barometer read 983 millibars, still falling. Through the revolving window on the bridge Keith saw his PhD project start to disintegrate. His deep-sea camera and core pipes broke loose from the for'ard hatch. He saw two hefty Fijian crew members go forward, saw green seas wash the two giants into the scuppers, and knew there must be no second attempt. No such kit was worth a man's life.

The wind howled, the waves rose and the marine radio chattered with emergency responses to the *Wahine*'s grounding on

Barrett Reef. The great storm had begun picking apart the *Taranui* too. The small propeller under tow to record the ship's speed broke free, then the anemometer disintegrated, and then a wave carried away the starboard wing of the bridge, shorting out the ship's radio and the chart lights in a blue flash. Below, the ship's engine was overheating, sucking air, and the emergency klaxon came on and stayed on. Keith thought of his wife Val, and the gypsy-like prophecy of Val's mother when first shown the couple's engagement ring with its sea-green tourmaline set between two pearls — *You will be divided by water*. Someone turned on a transistor, and National Radio reported the *Wahine* broken loose from the reef, and drifting down-harbour. Keith watched the *Taranui* skipper write in his log, *Seas precipitous, 40 ft*, and then they were in the eye of the storm, the seas fell away, the klaxon fell silent, and the worst of it was over.

Keith stepped out on deck later. His seismic system was housed in a shed but the roof was gone. He opened the door. The big radio valves were broken, the wires tangled, and seawater poured out from two years of clever makeshift design. Keith was left with one rock dredge, and as the *Taranui* quartered the seabed off southern Hawke's Bay, he dutifully knocked mudstone off the rocky offshore ridges and dumped them into labelled sugar sacks.

The foraminifera fossils Keith gathered were pinhead size maybe, but just like the larger macrofossils of bones and teeth, the foraminifera had a fossil index. The delicate and distinctly different constructions of the foraminifera's inner shells were charted on a micro-fossil index that listed water depths for each distinctive type.

Later, Keith emptied his sugar sacks. He began analysis with a microscope. He saw foraminifera fossils dredged from the shallow ridges off Hawke's Bay that simply should not be there. They were remnants of ancient but utterly distinct deepwater organisms.

The index also charted foraminifera by age and, to add to the puzzle, the foraminifera on the shallowest ridges were older than those on the deeper ridges. That also should not be.

Keith stuck with the results and published them in an obscure journal, but they made no sense, and if his contemporaries noticed at all, they might have put it down to the inadequate techniques of a novice oceanographer.

Plate tectonics theory first gained its bold name in 1968, but it was still largely untested, still shaking free of its reputation as geopoetical speculation. The Swiss geologist Rudolf Trümpy would later make his own poetic pronouncement on it — 'Like Venus, the theory of plate tectonics is very beautiful and born out of the sea' — and in time it would change almost every facet of geology, but in 1968 no one in New Zealand, and probably too, had they been asked, no one even in world geology was yet close to figuring out what the peculiar composition of the ridges off the East Coast of New Zealand meant, and how they too were a glimpse of the newly emerging Venus.

In 1985, forensic geologists helped convict two of the French agents who'd sunk Greenpeace's *Rainbow Warrior* in Auckland harbour and drowned Fernando Pereira. Police had seized a rental van, and the DGSE agents inside were posing as Swiss tourists, but the grains of Tangihua basalt and the dust of micritic limestones in the van tied them to Northland, where two limpet mines had been unloaded from a New Caledonian yacht. McCallum red chip tied them to the footpath along Tamaki Drive, where the two DGSE frogmen had been despatched and retrieved.

The agents were convicted of manslaughter and sentenced to 10 years' jail. The French Government retaliated with trade sanctions. United Nations diplomats intervened to convert the ugly impasse into mutual concessions. New Zealand would allow the convicted agents to complete their jail time on a French atoll, and France would pay New Zealand $13 million in reparation.

If you'd modelled the Hikurangi Trough in soap, then washed your hands with it for a week, that pretty much described New Zealand's bathymetric maps of the early 1990s. Conventional sonar gave you bathymetric data in a single line — hill and valley relief as a child might draw. By the 1990s there were a lot of single lines, but they were still often tens of kilometres apart.

The French research vessel *L'Atalante* was commissioned in 1990, years after the *Rainbow Warrior* was sunk and the reparations were agreed, but part of that settlement had been an agreement for the DGSE agents to serve their sentence on the French Pacific atoll of Hau. The French had repatriated them early, and the dispute still simmered. The French ship was outfitted with a new and revolutionary swathe-mapping system. Its multiple sound scanners returned a reflection from the seabed 10 kilometres and more wide, and the returning sound was converted not just into bathymetric depth but also into texture. *L'Atalante* could, in effect, illuminate the seabed, and demand for the ship was high worldwide. Her running costs also ran high. New Zealand couldn't afford her, but the French were persuaded a New Zealand project called Geodynz might qualify for funding by France's own science funding agency. In 1993, that agency had 40 propositions before it for work in the south-west Pacific and France had its own colonial territories in the Pacific to impress. The agency had enough money to fund only two projects, but New Zealand also had one huge and unspoken advantage over its competitors.

November 1993, night-time, and the New Zealand team leader aboard *L'Atalante*, Dr Keith Lewis, wasn't used to the luxury of it. He was used to winching the seismic hydrophones out and winching them back, to decks awash with seawater and the tingle from the electric boomers even through your gumboots, to numb hands and heavy piston corers that rolled dangerously on the deck. Now he'd been served cordon bleu meals, and wine from Bordeaux — but forget any of that. Night-time made no difference to a ship that could see in the dark, and surrounded by the glow of monitors, he was watching bathymetric strips inching out of the thermal linescan printer even while the ship was still under way. Instant maps. The bland swirl of New Zealand's bathymetric maps was resolving here into high definition. He was staring at direct rubbings from the underworld. Some of it he knew or guessed already, other details left him agog.

Previous surveys by the Navy vessel HMNZS *Lachlan* had shown some kind of jumble on the trough floor. The swathes detailed the jumble. The continental slope had collapsed here and huge blocks five kilometres and more wide lay strewn across the trough's bland plain. One block was so big you blinked and rechecked the scale of it. Seventeen kilometres long and a kilometre high, yet another relic of some prodigious event, and Keith grabbed the French team leader, Jean-Yves Collot, and together they persuaded *L'Atalante*'s captain to move closer to East Cape. It took *L'Atalante* two hours to sail the length of a 40-kilometre collapse off the continental slope, and the debris flow away from that collapse would prove to be 170 kilometres wide, but strip by strip the ship mapped what Keith and Jean-Yves named the Ruatoria Debris Avalanche.

Later, the swathe-mapping data was ground-truthed with piston core material, augmented with data sets from earlier surveys, and what was by then the National Institute of Water and Atmospheric Research published its Geodynz map of the North Hikurangi Trench. The offshore was no longer a gnomic place where fish slunk about. The topography was as clear as the onshore and anyone who saw it made the conceptual jump. Onshore, offshore, it was a single system. New Zealand's rumpled surface was directly connected and controlled by the submarine architecture below.

Such was the revolution that Keith Lewis helped pioneer, by the high-resolution maps that resulted from Geodynz, and the series of scientific papers he co-authored with Jean-Yves Collot, and then he also did the entire country a favour. The articles he published in the *New Zealand Herald* and other newspapers set out vivid three-dimensional illustrations of the country's offshore landscape. The little grey bumps of the onshore topology looked like hovels beside the height and depth of the undersea, whose steep walls and arched apertures, flying buttresses and onlapping roof gave it the presence of a great church. I remembered those articles, remembered sitting at the kitchen table shaking out the height and width of the broadsheet *Herald* and gazing for the first time at New Zealand's underwater landscape. Sitting there, over coffee, you recognised grandeur. At one end of it was the Ruatoria Debris Avalanche, where the church looked to have been hit by a howitzer.

Keith was waiting for me under an oak tree with a flagon of craft beer. He lived with his partner Robbie du Chateau in a Victorian villa recently emplaced on their one-acre spread at Paraparaumu. The villa had come up for sale in Eketahuna, Robbie liked it, and Keith had the house uplifted onto a lowloader trailer and tractor unit, following it down through the night in his CRX Honda. Like any good scientist, he wanted to date the acquisition. He searched the Alexander Turnbull Library's photo collection and established the villa wasn't on its Eketahuna site in 1895. But it was there in 1902. Those were the dating constraints. Peer review of any geological paper would give you a fail if you didn't put in the constraints, but when it came to giving the villa its due dignity and status Keith cast scientific rigour to the winds. A leadlight stained-glass window now sat above the villa's entrance lintel. Keith had cut the glass, moulded the lead piping, and at the centre of his art nouveau swirls, he'd crafted the old-fashioned orange numerals of the villa's date — 1897.

I volunteered the logic of why I was there. I'd studied long trails overseas, and they usually followed a single geographical feature. The Appalachian Mountains on the USA's eastern coast had their namesake trail, the Sierra Nevada above the western coast had its Pacific Crest Trail, England's central massif had the Pennine Way.

I told him what I wanted. Te Araroa didn't follow a single geographical feature. It was a mash-up of almost every landscape you could mention. The trail's unity, I'd always claimed, was deeper than geography, it was tectonic. It was two plates contending, a tug from below. I wanted stories from the subduction zone.

Keith recalled his first gut surprise at New Zealand tectonics. As a graduate of Reading University with an MSc in sedimentology, he'd emigrated to New Zealand to join what was then the New Zealand Oceanographic Institute, and within days of his arrival he toured the lower North Island.

— I can date that pretty good, 22 November 1963. I left Wellington with Jim Brodie, the institute's director, and we stopped for morning tea at Foxton. That's when we heard President Kennedy had been shot.

They walked on the volcanoes, paused at the hot pools, and came back through the Wairarapa. The NZOI director stopped the car beside one of the Te Aute limestones, and the two of them climbed a hill to reach the outcrop. Brodie turned to his young recruit.

'What age do you think this is?'

The answer was obvious enough to Keith. The outcrop was similar to the ones he'd studied on the chalk downs outside Reading, right down to the brachiopod fossil he could see poking out.

'Cretaceous, probably over 100 million years old.'

'No, it's lower Pleistocene, less than two million.'

— That was mind-boggling, said Keith. We were standing on a serious hill, and that rate of uplift was beyond belief. You've got to realise the Pleistocene stuff in north-west Europe is either glacial debris, or a smear of mud in the bottom of the Thames Valley — that's all it is.

Keith poured another two glasses of Emerson's Bookbinder Draft, and I listened some more. The Geodynz story. The speculation on what caused the huge collapse off the coast of Ruatoria. The seamount that had disappeared. Good stories. I thought I wanted to get him out onto the road. Geologists are busy people and not everyone could spare the time, but the housetruck parked in the yard was a good sign.

Later, we took carrots from the kitchen and went on down to the lake at the bottom of the big lawn. The lake had an island in the middle where a rabbit had dug sufficient burrows to qualify as a warren.

We threw the carrots across.

— You believe the Hikurangi Trough swallowed a seamount?'
— Perhaps.
— And that's why the East Cape is rising as fast as it is?
— Yes.
— I'd like to go with you, I said, out to the Cape, and tread that rising ground.

Keith thought about it a long while.

— Okay, but if you want to understand subduction, we should start from Auckland.

We came over the Auckland Harbour Bridge, and Keith gestured west to the distant ranges.

— The Waitakeres. Part of the first magmatic arc.

We left the southern motorway, cutting through to Clevedon and on over the Kawakawa Bay road until we surmounted the last hill, and the Firth of Thames lay open to view. On the far side of the Firth, the blue summit humps of the Coromandel Peninsula rose through a delicate layer of cloud.

— Coromandel. Part of the second magmatic arc.

We stopped at Te Puke to hunt for a $2 shop. Keith bought a folding fan. We strolled down the main street looking for a café, took delivery of our flat whites and sat down at a table.

— This first rib is the Waitakeres. They're Miocene, the magmatic arc, say, 20 million years ago.

Keith began to open the fan.

— The Pacific Plate boundary is rolling back clockwise, like this. Now this second rib is the Coromandel, the next magmatic arc. That's Late Miocene to Pliocene, say 10 million years ago. We're still opening up here. The Northland Allochthon is separating out from the East Cape Allochthon. Then this third rib, that's us. The volcanics start in the Quaternary, say two million years ago, the Taupo Volcanic Zone. It's the magmatic arc from the subduction that's under way off the East Coast right now, and we'll be driving across it.

We followed State Highway 2 towards Whakatane. Volcanic Whale Island stood offshore, and beyond it the steaming summit of White Island. We passed alongside seaside cliffs almost purely white.

— The casual visitor might think this is limestone, said Keith. In fact it's volcanic ash.

Beneath the white beds lay deeper, coarser and darker beds where ignimbrite from the long-ago Okataina caldera had hit the sea and exploded. Evening came on, and we turned inland off the highway towards Kawerau. Putauaki rose in front as a simple dark cone against the sky. We hired a cabin and spent half an hour bathing in the hot springs that bubbled up along the Awakeri faultline, watching steam rise through the light and vanish into the trees overhead.

Next day we came on through Opotiki. It was the gateway town to the East Cape, and part of the North Island's grid of Maori towns. As we drove on east a sign said *Farewell and Good*

Luck. That seemed right. The East Cape was frontier country. The Raukokore Church was a frontier church, standing alone on a coastal promontory with two live horses recumbent at the gate, and penguins nesting under the altar. The horseman holding the rope bridle was a frontiersman who cantered bareback along a beach. The signs along the highway posted frontier warnings — *No Shooting Stock* — and bugger off messages — *Stop Deep Sea Oil. No Drilling Spilling. Danger Fracking and Radioactive Contaminants.*

In June 2010, the Brazilian company Petrobras had taken up a five-year exploration permit over the offshore Raukumara Basin. Its ship *Orient Explorer* was running seismic lines across the basin in April 2011 when a protest flotilla of five vessels forced it to heave to. The company surrendered its permit in December 2012 without specifying its reasons.

We drove on. A team of contractors was threading through the new blue broadband cable to tether the East Cape back to New Zealand, but there was no guarantee the Cape wanted that, and every hilltop seemed to have a white horse, watching.

— Allochthon to the left of us, allochthon to the right, murmured Keith as we left the coast and started to go cross-country to Hicks Bay and Te Araroa township.

It was the kind of stratigraphy I'd seen before: basalt summits, rootless, and soft layers under. The same bands in the same sequence. The Northland and the East Cape allochthons were a single event, slanting in from the north-east, but they'd been separated as the magmatic arc jumped from the Waitakeres, to the Coromandel, to the Taupo Volcanic Zone. The arcs jumped, and the intervening basins opened. Northland had gone quiet as the trench moved away, but the East Cape had suffered double jeopardy. The original soft layers of its autochthonous country rock were overlain by an allochthon that was itself mainly smashed and soft layers, and all of it subject to the continuing pressures of the subduction zone. Every outcrop was shattered, and the soft broken rocks were eroding, washing out everywhere.

We left Te Araroa, driving under Miocene cliffs set with oyster fossils that broke into meaningless shards when you tried to remove them. A heavy-headed heifer blocked our passage across a one-way bridge, and took her time to lumber clear. We stopped the car two kilometres beyond Te Araroa, walked out onto the coastal platform and patted the seep limestones, the natural pipes five metres and more in diameter, that once brought fluid up from the depths and whose tops stood proud of the mudstone platforms like doughnuts on a tray. The Cape was full of seeps, cold fossils like these or warm seeps bubbling in a farmer's field, or seeps out to sea, for there was a working petroleum system underground.

New Zealand's biggest known oil fields are in the west, on the landward side of the continental mass that rifted away from Gondwana and began to sink. The rivers became sluggish, flood plains broadened, water ponded, and anoxic marshes tamped their quintillions of waxy coastal plants into the mud.

If you looked to future prosperity in the only scrape of land that would survive the sinking in that western province, then the trick was in the wax. Taranaki's geology didn't favour oil. Any seismic survey there shows the sharp reflection from coal seams deep beneath the ocean floor, but the conventional geological wisdom would inform you also that such undersea coal seams are of little interest, because coal does not, by any known geological process, produce oil. Except that in New Zealand it did.

Edward Metcalf Smith, employed in 1860s New Zealand as armourer and small arms advisor to the New Zealand Field Forces at Fort Britomart Auckland, and later armourer to the Taranaki Militia and gunsmith in New Plymouth, had no geological wisdom and knew no constraints. His life and trade were tied to British steel and fine oil, and as an Englishman set down in a new land, he was a fierce advocate both for the iron sands along the Taranaki coast and for bottles of the fluid that oozed from Ngamotu Beach on the New Plymouth shoreline. He had useful British contacts, and when the Birmingham Chemical Association reported the bottled fluid was good-quality petroleum, Smith formed a syndicate and drilled one of the earliest oil wells in the Western world, trailing the wells of Ontario and Pennsylvania by six years, but neck and neck with the first wells drilled in Texas.

The syndicate's Alpha well, complete with its iconic derrick, yielded, at its height, two barrels of oil a day, and many investors lost their shirts then and for many years thereafter, but it would take a more modern technology to tell them they were right. Currently the region has 19 oil and gas fields producing something over 12 million barrels of oil a year. Coal seams from the Carboniferous period over 300 million years ago do not naturally produce oil, but Taranaki coal formed under deep sediment loads between 73 and 35 million years ago. Taranaki's coastal plants were evolved far beyond their Carboniferous forebears, their waxy cuticles reducing water loss and serving also to protect their leaves, flowers and berries from coastal salt and wind. Under the heat and pressure of deep burial, most of Taranaki coal's woody cellulose and lignin stayed underground as a residue, but let go its waxy paraffins and napthenes.

The other early promise of oil came from the seeps in and around the East Cape, and along the East Coast. The history echoed Taranaki's early entrepreneurial wild speculation and vain hopes. Syndicates drilled alongside seeps, but the industry was never commercial. Yet here, too, as in Taranaki, the early entrepreneurs weren't wrong.

The East Cape and East Coast formed along Gondwana's sunken Pacific margin. Unlike Taranaki, the oil-bearing strata were marine rocks rich with carbon from marine microfauna, but mixed too with debris from onshore vegetation. The marine strata, partly uplifted by subduction pressure, form the East Coast's highly visible Waipawa Black Shale and the Whangai shale. Onshore, those strata are immature, without the heat or pressure to crack the organic compounds into oil, but the strata then dip away offshore to underlie the Raukumara Basin.

Oil prediction is complex. It uses software and computers, but the two primary indicators are simple: temperature and pressure. For every kilometre of sedimentary depth add 20–25°C of temperature, and where the strata reaches four kilometres depth — where the heat in the oil kitchen is high enough to brew coffee, and holds steady for a million years — the complex molecules of hydrocarbons in the source rocks begin to break down, become smaller, simpler. They begin to liquefy. And then you have oil.

Oilmen drilling in search of that oil can predict the pressure produced at depth by the weight of overlying strata. They counter that pressure with drilling mud that goes down with the bit. But close to a subduction zone, the fluids in rock pores, already squeezed from the top, are squeezed from the side also, and the region can be overpressured in unpredictable ways. The East Coast has that reputation. For any well drilled here, the oilmen must adjust the weight and densities of their mud to compensate for overpressure, but the calculations are more complex. Get it wrong, and the well will blow out — its mud, and even its walls.

Oil software has fingered the Raukumara Basin nonetheless as a good prospect. It has the source rock and it has the kitchens. You could claim, as a confidential GNS document did in 2008, that the source rocks might have expelled up to four million barrels of oil per square kilometre since the Early Miocene. That paper did the maths — 'Assuming these volumes of petroleum are expelled over 10,000 km^2 an estimated 40 billion barrels of oil has been expelled from these formations in the Neogene.'

Expelled, but relevant only if the migrating oil has been caught and stored. Oilmen study the stratigraphy of their seismic shots to look for reservoir rock, the porous sandstone or fractured limestone filled like hard sponge with ancient seawater that's then displaced by the migrating oil. And they look for a wide dome above the porous reservoirs, an impermeable cap of mudstone to capture the oil that meanders ever upwards along fault planes or through the craquelure of subduction-fractured rock.

— There's plenty of places on the East Coast where it seems to be the right temperature, but there's no good capping in most places, said Keith. You need a nice mud on top of a porous limestone or sandstone. There are places where they think they can see it, but not over a big enough area. It's too small to spend that much money drilling it. The right combination could well be in the Raukumara Basin, I don't know. That might not be what Ngati Porou want to know.

We climbed the 750 steps to the East Cape lighthouse, and looked out to sea. The lighthouse, tall and white and masterful. The ocean, blue and flat. The lighthouse could sweep its beam across 35 kilometres of ocean. Undersea, that same light might make 50 metres, and be no more than a glimmer.

The flax waved in the wind and the manuka bobbed as we stood on the lighthouse knoll. New Zealanders would never see their most sublime landscape offshore, but Keith Lewis might offer some equivalent of a scented garden for the blind. Out there was the shallow slope of the continental shelf, and then the sudden three-kilometre drop-off to the trough floor. They called it the continental slope, but it was less a slope than a precipitous medley of sheer rock faces, and sediment-laden shelves made slick by layers of underwater clay — the slippery smectite, and bentonite. Its ridges and cliffs were oversteepened often by erosion, their mudstone or sandstones faces riddled with freshwater seeps, gas seeps, oil seeps and the cryogenic layers of frozen gas hydrates. All of it was shaken at its base by the subducting plate, yet even the inherent weakness and instability in this trench wall couldn't account for the size of the avalanche identified by *L'Atalante*.

— The Ruatoria Debris Avalanche, said Keith. Roughly equivalent to Coromandel Peninsula falling off the top of Mount Cook. We believe a large seamount began to impact the wall of the trough about two million years ago. The wall became unstable and avalanched maybe 170,000 years ago and, meantime, the seamount is still going down, and we've got the East Cape doming upwards fast, over four metres every thousand years.

On known angles and subduction rates, the seamount was currently some 15 kilometres below East Cape. It was possibly as big, judged by the height and width of the Ruatoria Debris Avalanche, as Ruapehu. The subducting plate had pulled it through the continental slope, and taken it down.

I suggested I should write to the New Zealand Geographic Board and propose the naming of Mt Lewis.

— Let us, replied Keith, not go there. I doubt that the Geographic Board would be naming features that haven't been geography for a million years.

The Ruatoria Hotel was friendly and kept faith with its recent past. We ascended the stairs and alongside us on the walls were the 1970s All Blacks. Andy Leslie was the arms-folded captain in a 1974 team that included Bruce Robertson and a young Joe Karam. Bryan Williams, Sid Going and Ian Kirkpatrick hung there, individually framed and slightly foxed.

The rooms were impeccably clean and tidy, and there were a lot of them. We made our way along a corridor, and left and right of us the doors stood open on rooms with candlewick bedspreads of baby blue, or cream, or orange, or plum, or pink, or pale green, or yellow. Every bed had an elaborately folded clean white towel and a quilt for cold nights. Reflected from every mirror on every dressing table in every room was a white plastic doily, fragile with age and as delicate as coral, and on it a tiny porcelain vase with a single fabric rosebud.

Keith was a beer connoisseur, and every night on the road we'd tried a different beer. This time he outdid himself, producing two bottles of Belgian lambic beer. The niche microflora within the traditional lambic brewing equipment is never entirely cleaned out. The hot wort of unmalted wheat and malted barley mash is cooled in vats under roofs open to the valley winds, and the wild yeasts, and the birds. Lambic beer production has remained almost unchanged over generations and assigns a value to decrepitude.

— Some say it's the birdshit that makes the difference, said Keith, taking the slow sips of a connoisseur.

We drank our lambic beer in the late afternoon and gazed out from the window. Birds flew in and out of the walls of the Ruatoria Hotel, the hotel's guttering grew its wild grasses, and the paint peeled, and in the morning we sat out in the private yard, eating breakfast in the sun. Beyond that high enclosure, Ruatoria was stirring. We heard the cars of the workers go by, then the steady clip-clop of a horse approaching, and above the blank fence we saw the head and shoulders of a Maori rider glide by.

Te Runanganui o Ngati Porou were custodians of Hikurangi, and with Ngati Porou guide Paora Brooking we set off for a close look at the East Coast's highest summit. We drove across the Waiapu River bridge, and the river below ran in quiet channels through a wide bed of shingle.

In 2003, NIWA produced a map showing how much of New Zealand was being washed out to sea, and by what rivers. The Waikato offloaded 370,000 tonnes a year, the Clutha 390,000 tonnes, and the Waiapu River 35 million tonnes. The Waiapu sediment density was so high that it was perhaps best to look for comparisons in the wider world. The Mississippi's notoriously muddy water carried off 150 million tonnes a year from its lower alluvial valley, raising the bed and expanding the deltas. The Waiapu's sediment load was 23 per cent of the Mississippi's load, but the Waiapu catchment was only a thirtieth the size. Cubic metre for cubic metre, the Waipu was carrying more than seven times the Mississippi's load of fine sediment.

We turned off the highway into the Tapuaeroa Valley. The road stretched on through a typical farming valley, with sheep and beef cattle and half-round barns in the fields, the low-hipped roofs of New Zealand farmhouses with trumpeting pink flowers of naked lady at the gate. The road crossed one-way bridges over watercourses that were less like streams than shingle slides coming off the low hills. Then the valley opened up into depth and distance, and five fearsome maunga. Each one rose in straight walls of rock above a green skirt of farmland. Each summit was a solid stone block with sheer sides, vertiginous hideouts, and burials in its nooks and crannies. The biggest maunga among them was Hikurangi, the first part of Te Ika a Maui to surface on Maui's bone hook. Paora told the stories well, and it was no time for geology.

— You can see this land was pulled up from under the sea, Paora said, and he was right. It was also true that too much of it was going back to the sea. The valley narrowed, and the road began to run alongside the wide Waiapu.

— Every year, the river eats land, he said. Our old people say their old people could jump across this river. But since the trees were cut, that turned the erosion on. Half the valley has gone. There used to be twelve football fields of flats on the far side. Twenty years ago they logged the trees and the land has been moving ever since.

We crossed the Waiapu again on a bridge built on top of an unseen, buried bridge. Bonnet slanted to the sky, we drove up

Hikurangi's great flank on a farm track, past sheep and the prone trunks of a long-ago native forest. At 1000 metres, we came to the sculptures that stood on a flat semicircle of land. They told Maui's story. The enchanted jawbone. Later, the demi-god's bold venture to secure immortality for humankind by slipping up through the obsidian-laced vagina of Hine-nui-te-po. The carvings stood as guardians of myth at the foot of Hikurangi's own summit walls, its rock chimneys, and tumbled blocks of sandstone and the petrified parts of Maui's canoe.

The measurement of a river's coarse sediments is trickier than measuring fines from a test tube, but Keith Lewis had his own rule of thumb. Once back on the river flats, he stepped down from the 4WD. He was looking for the farmhouse. In 2001, he'd photographed the Waiapu running its shingle through the farmhouse kitchen. In 2007, he'd photographed the river running its shingle halfway up the windows. Keith knew where the house had been, took his fix from surrounding landmarks, spotted what he was after and photographed the last of the sequence, the remnant top of the farmhouse chimney, grey concrete hugged by the grey encompassing shingle.

The river was as broad as a South Island braided river, a wide conveyor abrading the hills and dissolving the valleys. Every side stream was a shingle slide subsiding slowly into the main flow, and it was distressing to look at, but you made of it what you could. On the return journey Paora stopped on one of the side bridges, and we looked up the shingle slide to a large boulder under transport by the creeping mass.

— That boulder. When it gets down this far, I want to take it out and use it as a memorial to my father, Jack Brooking. He's one of the best sculptors on the coast, and I've told him, 'You're going to have a big rock, and all your mokopuna are going to come and sit on it. And I'm going to have a pair of hands put on it. Your sculptor's hands.' He was born without thumbs, and that was a sign.

We drove on towards Gisborne. It was a big blue day, and big blue markers stood beside the highway, keeping the tourists on track. The road was not just a road, it was the Pacific Coast Highway, and the blue signs bore the highway's logo, a big white wave curving down, its leading edge curling up. It looked like a tsunami warning but, then, the big wave was on our minds. Twenty kilometres out from Gisborne we came down from the brown hills, within sight of the coast again, and crossed the Pouawa River on a perfectly ordinary two-way concrete bridge, except that on 26 March 1947 its predecessor, a 16-metre wooden span bridge, was lifted off its foundations and carried 800 metres inland. The same wave left seaweed dangling in telegraph wires and fish gasping far inland. Detailed eyewitness accounts of the 1947 tsunami are hard to find, and I thank Alan Tunnicliffe, who published his father's autobiography in a limited edition in 1990, from which the following account is drawn.

Don Tunnicliffe had piloted a Beaufighter on 47 missions during World War II, had been wounded by ack-ack shrapnel and awarded the DFC, and had returned to Invercargill postwar to wed his sweetheart, Novena Wensley. In March 1947, the two were three weeks into their honeymoon, exploring New Zealand on an ex-serviceman's free travel warrant and calling on Don's war buddies.

Fred Hall was a close friend from 489 Squadron RNZAF, and after enquiring at their Gisborne hotel where Fred might live, the two honeymooners set out by service car at 8 am, stopped at Turihaua Point, stepped down through an opening in the barbed-wire fence that bordered the road, and knocked on the door of the cottage below. They'd expected Fred, and found instead they'd arrived at the fishing bach of Fred's parents. The Halls welcomed them in anyway, and offered a breakfast of bacon and eggs. The two honeymooners sat at the table, and the Halls tended the stove at the other end of the room. Someone was riding in from the north on a powerful motorbike, and Don leaned towards the open door for a view of the road. It was empty. The elderly couple at the stove showed no alarm, but still unseen, the motorbike was opening right up.

Don went to the door and stepped outside. He saw a wall of dirty water risen 10 metres above the bach, boiling and curling. He saw a young man squatting on the concrete path outside, engrossed in pumping a Primus stove. He yelled 'Look out!' but saw the Primus youth whisked past, and then a seething mass of sand and seaweed tumbled Don up the beach into the barbed-wire fence and held him there under a press of black water. He kicked and twisted with the frenzy of fear, gouging his leg and tearing a clean strip of trousers from the knee down and right through the cuff, but he came free, and the sea took him on and dumped him on the road. He stood upright a moment and another wave came from behind and knocked him flat. He stood up again to see the first surge that had carried on uphill was now bearing its flotsam back down. The returning surge met more waves coming in and the beach below became a maelstrom of debris that popped the beachside smokehouses and storage sheds into the air with a noise like artillery and smashed them down in pieces.

The young Primus man picked himself off the hillside and came towards him distraught, and from their high vantage the two of them watched the Halls' bach being destroyed. The side bedroom went. The roof peak bobbed around, the lean-to boathouse and its launching ramp were ripped away and smashed. The kitchen held, but the maelstrom continued to surge and charge, stripped away its weatherboards, plucked out its bentwood chairs. It was a steady demolition, and Don Tunnicliffe looked down at it, stupefied, without sight of his new wife, and thought — *It's like watching murder!*

An arm appeared through the kitchen wall. The quavering voice of Hall senior called out, 'We're alright,' and Don Tunnicliffe waded through four feet of muddy water to lead the kitchen trio up onto a bank above the road. Don went back down and found woollen coats still hanging in a stranded wardrobe, and then they stood there, arms hanging — 'five wrecks of humanity' — until a jeep appeared, the driver hurling debris off the road as he came on. The driver picked them up and took them back to Gisborne.

We drove around the Turihaua Point and on to Tatapouri. A sign cut in the shape of a stingray advertised a beachfront stingray-feeding enterprise. A white painted crayfish logo nearby advertised the Tatapouri Motor Camp. It was, said the sign, an eco-friendly beachside paradise.

We turned into the campground and the manager, Chris Berge, was taking time out for a cup of tea.

— Yeah, that's my name but everyone calls me Budge.

He stood one level above us on his verandah, and we had a shouted conversation. We wanted to locate the site of old Tatapouri Hotel, and he pointed out where it had been, a few metres above the high tide mark, no more now than a line of kerbing. On that same March day, 1947, the publican's daughter June McLauchlan had noticed the tide edging the front lawn and had run inside to tell her father and mother. They'd all had time to scoot up the hill behind then turn to watch the incoming wave wash through the hotel and demolish a shed and a cottage alongside.

— Born and bred here, shouted Budge, and since I've been alive I've had about a dozen tsunami warnings. We're in the shit mate. Right here we're absolutely in the firing line. More so than Gisborne city, way more so. It's got a headland that protects it from a southbound tidal wave. See that blue drum sitting in the middle of the channel? At dead low tide I can put my feet on the bottom of the ocean and I can't touch that drum, it's above my head. The last tsunami, about three years ago, that drum was bouncing off the seabed. The sea sucked out that far. That was a one-off, an earthquake off Chile or something like that.

The camp had an evacuation procedure, and the Gisborne District Council had just sent him a revised run-up height for his cabins from 30 metres elevation to 50, but it didn't worry him. The cabins were already above that height. Chile was a regular tsunami source, but even the largest of them, the 4.5-metre wave that hit the New Zealand coast in 1960, was not devastating. And the typical time lag before it hit was 12 hours. Any local quake would allow only minutes of warning.

— Just a localised slump off the continental shelf, eh? Budge asked. He had an expert to hand, and he was interested.

— Yeah, well, you've got the Poverty Bay Indentation out there, said Keith. The northern edge of the re-entrant out there is oversteepened. Probably a seamount impact, about twenty kilometres out.

Everyone looked out onto the blank ocean, to the sun, and the salt, and the rock platforms. It was low tide, but offshore the stingrays were waiting for their free feed, and the crayfish, if Budge had it right, were waiting under every ledge.

— The localised ones, we've got seconds to react, right? said Budge. But the earthquake will be significant enough that you'll have real trouble standing on your feet. Evacuation, yeah. Over the road and up the hill as fast as my legs will carry me.

The disturbing thing about the wave of 26 March 1947, though, was that few people noticed the earthquake. On the Richter scale it registered as magnitude 5.8 and on the Mercalli scale of shaking it came in as a IV — a vibration like heavy trucks passing, a rattling in the windows, a creaking in the walls, but that was all. The puzzle never went away. In 1982 the *Royal Society Journal* published an analysis by the seismologist G. A. Eiby, who noted the two large gas-driven mud volcanoes in Gisborne's past. He noted eyewitnesses who'd seen large patches of ocean turmoiling white as the big March wave came visiting, and suggested in his conclusion that a moderate undersea earthquake might have triggered another gas eruption, and the gas eruption the wave.

He was shooting in the dark, working with the Oceanographic Institute's undersea maps of his era. He could not see what would later be called, with the understatement and precision that marks out geology's definitions, the asperities on the ocean floor.

In 2008, a report on East Coast tsunami hazard noted that the GPS network data over 12 years showed the converging plates off Gisborne and Napier were shedding most of their elastic strain. Fluids and sediment lubricated the interface between the plates, and there was little to suggest an earthquake could nucleate within the shallow reaches of the zone. Except that, in nature's perverse way, the 1947 earthquake had done just that. The report suggested an asperity as the likely cause, specifically, a subducting seamount, and noted that any such earthquake might rupture outward from its hard-rock epicentre through the ponded surround of weakened rock and saturated sediments. If they did that, the rupture would be slow.

In 2013 a team led by Rebecca Bell from Imperial College London set out to reverse-engineer the earthquake and its wave. The team had the original thermal paper image from the Wood-Anderson seismometer that recorded the 1947 event from a distance of 150 kilometres. It showed a peak amplitude of 70 millimetres. That translated to a reading of 5.9 on the Richter scale but the Richter scale didn't factor in time. The 1947 quake had gone on for four minutes.

Since the late 1970s, seismologists have favoured classifying earthquakes by their Moment Magnitude — a calculation of energy release that factors in fault length and displacement alongside seismographic evidence. For most earthquakes, magnitude measurement by the Richter method or by Moment Magnitude gives a similar result, but for a slow earthquake, the two diverge significantly. The 2013 team reaffirmed a startling estimate first put forward by GNS seismologist Gaye Downes in 2001. The offshore Poverty Bay quake was a magnitude 7.

The team also took note of pre-existing seismic data that had illuminated the faint bump of a stranger 10 kilometres in and 10 kilometres down on the plate interface. Their new magnetic survey confirmed the target, suggesting a lode of more magnetic rock down deep, a bump pressed downward and held as you might press down a beachball. The deep target coincided with the epicentre as calculated by 1947 seismograms.

Here, reasoned the 2013 team, in a region where the huge incoming plate and its slow yearly shunt was rendered slippery and aseismic, still they'd found a single regal asperity that might, within its own domain, hold back the tide more successfully than Canute. But only for so long. The convergence here was reckoned at 55 millimetres a year. That was 5.5 metres every hundred years, a stress beyond the capacity even of a mountain.

The team set out to synthesise the oscillation of ground motion and acceleration and duration as had been recorded on the long-ago photosensitive paper of the Wood-Anderson machine. They calculated the rupture plane as 55 kilometres long and 50 kilometres wide, and across that wide plane, the best match for the Wood-Anderson oscillations came at rupture speeds between 540 and 1080 kilometres an hour. What sounded fast to a layperson was, to seismologists dealing usually in earthquake rupture speeds of around 7000 kilometres an hour, joyously slow.

Joyous because slow earthquakes increase tsunami run-up heights, and the figures they'd produced on their synthetic Wood-Anderson, when fed through a tsunami modelling program with the relevant bathymetric data, produced for the first time wave heights to match those reported by Don Tunnicliffe and the other witnesses of 1947.

Puzzle solved as to why coastal Poverty Bay got a big wave from a barely noticeable quake. The earthquake that produced the huge Indian Ocean tsunami of Boxing Day 2004 was magnitude 9.1, and the quake that produced the devastating Japanese tsunami of 11 March 2011 was magnitude 9. The tsunamis that came away from those deep ocean fractures were moving as fast as a passenger jet, but the speeds were still an order of magnitude slower than the whiplash rupture below.

The size of the Poverty Bay quake of 1947 was magnitude 7, a hundred times smaller than the generating earthquakes of the 2004 Boxing Day and 2011 Japan tsunami. Had it ruptured at the usual speed the quake might have produced a one-metre wave. But in a slow earthquake, the wave above remained keyed to the rupture below. Rupturing rock and rising water chimed together, and the wave, now stacked upon itself, began to prosper and became New Zealand's largest historical wave, demolishing seaside shacks, bowling bentwood chairs down the beach like tumbleweeds, and delivering fish to the Tatapouri bar.

We headed out of Gisborne, and offshore the character of the Hikurangi Trough changed from the rasp of subducting seamounts and their avalanches to a broad flat plain. If you extracted a stratigraphic column from this bland undersea platform, the black basalt of ocean plate would lie at its base, then the mudstone, then the fine-grained sediment of onshore erosion swept out by the onshore rivers. All of it inching south-west, year by year, under the Australian Plate.

Incised into this smooth plain was a channel. Keith Lewis glimpsed it first in the 1960s during a survey undertaken by the Oceanographic Institute vessel *Tangaroa*, but the soundings were isolated, the picture unclear. By the 1980s an extraordinary picture had emerged. If you drained the Mississippi, you'd fetch up the same kind of wide channel that, in New Zealand, lay two kilometres below the ocean surface. It was often five kilometres bank to bank, with meanders, with levees, with point bars on the inside of its wending curves, and it ran fast. You could see the tangential fling of material where it over-topped the levee on its sharper turns.

Keith already knew the source of the undersea river, but hadn't mapped it, and when *L'Atalante* finished its East Coast survey with a day to spare, he suggested the ship head for the Kaikoura Canyon. The canyon hosted whole layers of sealife, a crawling layer of blind bottomfeeders and worms, of fish feeding on the crawling layer, of big squid and sharks feeding on the fish, and on and up to its population of spouting whales, and beyond that the tourism enterprises that followed the whales.

L'Atalante entered the tourist zone, its multi-beam acoustic imagery systems blaring at a high-frequency 13 kilohertz. It came in like a bat out of hell, the whales fled, and Keith would spend some time afterwards assuring anyone who'd listen that the scaring of the whales was no more serious than clapping your hands to shoo blackbirds off the lawn.

Meantime he got the measure of the canyon. It cleaved the offshore shelf almost in two, beginning only a few hundred metres offshore, a 40-kilometre gash across the shelf that slanted down to a depth of 2000 metres. The inner edges of the upper canyon trapped longshore drift coming north from the South Island rivers. Those edges grew heavy with millions of tonnes of sediment over years, and they overhung a sudden drop to the canyon floor.

I'd brought with me for the trip a DVD of Richard Hammond's *Journey to the Bottom of the Ocean* series, and we'd played it one night at a motel. A predominantly blue planet floated inside a giant hangar, and thanks to CGI it could perform a number of tricks. It was so big, Hammond had to be uplifted by cherry picker to persuade the blue parts of it to perform. He suggested we should drain the oceans to look at the seabed, beckoned the camera into close-up, and obligingly the oceans did drain away, withdrawing from the continental shelf, cascading down the continental slope and retreating across abyssal plains to finally lay bare even the mid-ocean ridges. For a man who was gesturing at empty space, Hammond delivered the lines pretty well. He had his script, and the CGI wizards had factored in the planet and its requisite close-ups later.

The documentary used live footage too, and Hammond visited the Monterey Canyon, on America's west coast. Much as he'd clamped himself into dragsters in his better-known *Top Gear* role, the clear plastic cockpit of a two-man submersible was clamped over his head, and he descended into the undersea canyon. The visuals weren't great. The light disappeared, a neon-bright squid squirted on through, and then the craft's puny light beam showed nothing more than thousands of white motes. The submersible got to the canyon and threw a small puddle of light against the wall. I noticed Keith wince.

Later, when he was describing the Hikurangi Channel, and the kind of flows that rolled through it, Keith brought up on his computer a picture of an Australian dust storm. It had the same height and rounded front as a dusty Uluru, engulfing the land, the houses, and any city in its path.

— This is what I imagined when they went down in that little sub in the Monterey Canyon, said Keith. The canyons get small avalanches every year or two. Bits fall off the side and they're small enough that they stop and don't make that click into autosuspension. But even those would be enough to wipe out your little plastic bubble. I'd prefer the Monterey one where you can drive yourself. The virtual one, onshore.

The Kaikoura Canyon sloped to its greatest depth of 2000 metres, turned north and led away into the Hikurangi Channel.

The transition between the two was marked by a ripple of big coarse dunes, not unlike the ripples a retreating tide might leave on flat sand, except these dunes were 10 metres high and four kilometres wide.

By Keith's estimate it took a magnitude 7 or 8 earthquake for the Hikurangi Channel to run. The lips of the Kaikoura Canyon extended 40 kilometres, and under severe shaking hundreds of millions of tonnes of South Island sediment would avalanche off those laden lips and begin to roll down the canyon floor.

Quite when those big avalanches made the click into auto-suspension no one knew. They did know it occurred, and when it did, what had been a relatively sedate 40 kilometre per hour avalanche down the canyon changed character. It became the Hikurangi River in spate, dropped its heavier and coarse-grained sediments, and a grey storm sufficient to overshadow any city got up on its toes. Wide as the Mississippi, the turbulence of the massive Hikurangi River picked up more of the fine silts in front than it was dropping in its wake, and so kept gaining mass. The greater the mass the faster it moved, until it was rolling along at motorway speeds, and given even the slightest slope it could go on for ever, the whirling sand at its base eroding the channel as it went.

Keith had mapped the channel along the Hikurangi Trough until the flow was blocked by seamounts and avalanche debris, and it turned almost a right angle to carry on eastward across the Hikurangi Plateau. There, over untold millenia, it had cut itself narrow gorges 500 metres deep. At the end of the plateau, the Hikurangi River met the huge Deep Western Boundary Current, which carried away some of the riverine sediment, but the bulk of it still kept on, to end finally on the abyssal plains in a delta of tributary channels. By then the enormous density current had travelled 2000 kilometres. In an article for the *New Zealand Herald* Keith called it New Zealand's longest river and he'd reached for the poetry of Coleridge to capture the wonder of it —

Where Alph, the sacred river, ran
Through caverns measureless to man
Down to a sunless sea

'And perhaps what is most amazing,' he wrote, 'is that it doesn't merge with the huge ocean current it flows into as a tributary does when it meets a main river, but, incredibly, passes right through it to emerge on the other side. It's like two trains colliding and passing through each other without harm.'

We drove up Te Mata Peak in the late afternoon, parked and went on up to the trig. Havelock North and Hastings stretched out below, and Napier was an art deco smudge out to the north. Children squealed in the clear air. A line of what looked to be birthday party kids wavered in precarious poses along a row of parking lot posts until the grown-up at the end of the row got the right camera angle, took the shot, and one by one the kids dropped off, crouched and ran away.

We were at the highest point of Te Mata's limestone ridge. It dropped away on its eastern sea-facing side and the local hang-gliders had built themselves a launch platform there. The sun was dipping west, and Te Mata threw a long shadow over the hang-gliding arena. The slanting light threw into relief the serried ranks of high hills that stretched away inland, their scarps deeply shadowed, their rounded western slopes sunlit still.

— There's no way, is there, said Keith, you'd draw this other than as a scarp and a dip.

We laid out the maps, and looked at the undersea relief, the same ridges Keith had set out to study through the Wahine storm of 1968.

— We thought those ridges were ancient landslides, said Keith. We believed then it was a geosyncline that was going down offshore, and up onshore, and that the undersea ridges were slumps off the tilting slope of onshore land.

In 1968, even five years after the first beginnings of plate tectonics theory, the Old Geology still prevailed. Its commonsense explanation for coastal heights like Te Mata was that an offshore basin, the geosyncline, sagged under its burden of sediment, pressured the mantle, and the mantle, by the principle of isostasy, rebounded like a trampoline. The uplift out of the sea was obvious enough. Te Mata was full of fossil shells. Its uplift had the logic of a seesaw, and if you had an offshore synclinal sink and an onshore anticlinal lift, then it made total sense that any undersea ridges leading down to the syncline came from the anticlinal child sliding down the high side of the seesaw, not some clawing ascent by the synclinal child. The slide downward had gravity on its side, it had common sense on its side, and the ridges were slumps coming off the land.

Later you might stroke your chin and agree that this kind of local seesaw, even if it did take five million and more years, was unworthy of a great, and living, and generous planet, but in 1968 you didn't know that. There was only the prevailing common sense, and the commonsense conclusion that the oldest slumps would be those furthest out and deepest. The newest slumps would be those closest in and shallowest. That was the common sense of 1968, and then along came Keith Lewis in the MV *Taranui*, surviving the *Wahine* storm and trawling the ridges with the single tool left to him, the rock dredge.

Keith's strange findings: the shallow ridges — or call them, by convention of the day, the most recent slumps off the land — had deepwater foraminifera microfossils aboard. They also had the oldest microfossils. Those ridges far out to sea had younger microfossils.

That was what he found, and then he got on with other things, but the odd finding stayed with him over years, as a burr in the back of his mind.

In the mid-1970s he saw a paper by two geologists, one from Cornell University and the other from the Scripps Institution of Oceanography, headed 'Subduction and Accretion in Trenches'.

'Although the reality of subduction has been greatly strengthened by recent investigations, there is little information dealing with the mechanisms by which material is subducted or accreted to the upper plate,' the paper began. 'An attempt to determine the gross evolution of subduction zones has been made . . .'

The paper went on to discuss the deformation associated with young trenches, specifically the pile of faulted, ridged material, tens of kilometres wide, sandwiched between the underlying plate's entry and the top of the continental slope.

The authors had drawn a simple diagram that showed the subducting plate sliding at a low angle beneath the toe of the continental slope. The material coming off the plate was being piled against the slope in a series of ridges, and the whole pile was ascending.

'This region is a rising tectonic element,' said the paper. '. . . an accretionary prism develops . . .'

— You don't have many eureka moments in your life, but that was one of them, said Keith. That's it! Those ridges weren't slumping off the land, they were rising! They were coming up from the sea.

Everything made a sudden sense. It was mid-afternoon. He burst out of his office, and a biologist was walking by. The only person around, off for her 3 pm smoko break. He buttonholed her, sat her down, talked the revelation through in all its detail and sometime later she reminded him it was way past 5 pm and that they both had homes to go to. He realised he'd been talking his head off.

It was 1975. The Old Geology was being dislodged by plate tectonics, but only slowly. Keith gave his first talk on New Zealand's accretionary prism to a geology conference, heard the silence and retreated to the pub early, but later they crowded around and clapped him on the back.

— That was a great talk, Keith.

We were standing on Te Mata Peak, and Keith was pointing to the geological map, spread out, its edges flapping in the wind.

— You can see the boundary between what was scrunched up and true accretionary stuff that was actually scraped off the downgoing plate. You don't have that on land, but even seeing the top of that imbricate thrust zone like what's behind us here is pretty unusual worldwide. That's usually well offshore.

The submarine push had begun in the Oligocene. The land heaved up in overlapping waves, some of them undersea, some of them reaching onshore, and the process was rather like a house roofer doing his job by starting at the guttering and pushing all the tiles out from there. He takes his time, and the oldest tile will bear the signs of great age by the time it gets to the ridge line. Not an efficient working method, but without opposing thumbs to grasp, nor legs to clamber, the planet was nonetheless using its own abundant energy to rebuild its eroding continents.

Keith gestured at the ocean where the slowest onlapping tide of all was making its way gradually towards us.

— It's making new land out there, he said. Driven by the Pacific Plate pushing west. Continent in the making.

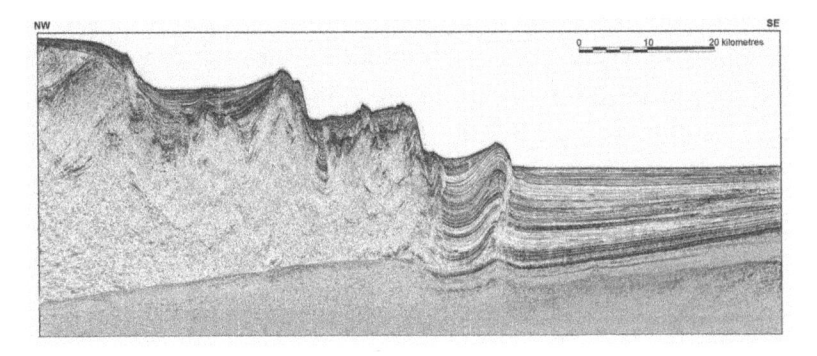

Golf was invented in Scotland, on the glacial topography left by the ice age. The St Andrews links and many other famous courses have undulant fairways rolled by the huge pressure of ice. The retreating glaciers have left behind pretty kettle lakes, errant rocks, or smooth hummocks that rupture into natural bunkers of glacial sand. Run-off from the glacier's melt has further shaped the outwash plains, and opened the gullies and the snaking eskers that separate fairway from fairway.

Cue Cape Kidnappers, and its marine terraces. We were parked on the 6000-year-old marine terrace at the Cape's base, making plans to get to the 120,000-year-old terrace. The problem was the golf course. In the New Zealand manner, one of the country's significant landscapes had been given over to sheep for most of its post-colonial life, then, in the New Zealand manner, sold to an American billionaire, Julian Robertson, who turned a large part of it into a golf course.

The golf course was gated and, aside from tradespeople and guests at the lodge, was usually open only to paying clients, but we rang ahead to the golf shop and got permission anyway, falling in behind a Liquorland truck then following around beside a stream and a big soft cliff packed with river stones.

Keith had come up to the Cape in the 1960s when estimating rates of uplift on the East Coast. The method was simple. Geologists had agreed worldwide on the dates of ice age glacial maxima when ice sheets and glaciers advanced and sea levels dropped, and interglacial maxima when ice sheets and glaciers retreated and the sea level rose. In tectonically active landscapes, the wave-planed lands of the interglacial maxima were uplifted into distinct terraces. Take two terraces, measure the uplift between them, apply the known dates of the two separate interglacial maxima that produced the terraces, and you had uplift rates.

It was rough and ready but in those days all the measurements were rough, and the young Keith Lewis was lucky to have with him one of the roughest geologists of all, the then 60-year-old Harold Wellman. It was Wellman whose blunt humour driving up from Wellington offered him the only instruction you'd ever need for driving the Land Rover — first and third gear on Tuesdays, Thursdays and Saturdays, second and top gear on Mondays, Wednesdays, Fridays and Sundays. It was Wellman who sent the same Land Rover crashing over an East Coast dune so fast that the altimeter popped off the dashboard and broke, and who turned to his young partner with a grin — *Now, we'll just have to boil water to get our height measurements.* It was Wellman on the road back to Wellington who'd stop to shovel up roadkill for the potato patch he'd famously dug into the capital's green belt. It was Wellman with the bold phrases on rock that you never forgot — *If it's old it's hard, if it's young it's soft.* It was Wellman, the man who'd discovered the Alpine Fault and insisted on its astonishing strike-slip movement, who'd come closest to intuiting the vast scale of New Zealand tectonics. By reputation he was abrasive and argumentative, but he got on well with Keith, and maybe the bond was Keith's excitement, newly arrived from the stasis of English and Welsh tectonics, at the prodigious heaving of the New Zealand turf.

The road curved up through pine forest, and we topped out finally onto a high planar surface and drove towards the golf shop. The flat surface was cut by gullies here and there, but they hardly dented the straight green edge that ran parallel to the blue sea horizon.

— Whoa! There you go! said Keith, looking around at the smooth surfaces. This is pretty good. We're on the 120,000-year terrace.

He squinted through the manuka, searching uphill for evidence of a previous interglacial maxima.

— Yes! Up there you can see the 240,000-year terrace. It's more deeply eroded, but you'll see the concordance where the hills chop off at roughly the same height.

New golf courses often need extensive earthworks as they strive to recreate the traditional landscape, but here there'd been no need. The undulant fairways of St Andrews had been rolled out with great force by the Emperor Ice, but the great force of Tangaroa had rolled out the undulant fairways at Cape Kidnappers. We stood on a natural plateau, and the fairways stretched out in front, flat fingers of Disney green pointed to the sea.

Golf also likes the clarity of an uncluttered background, and here too you noted the simplicity of the upright pin, the fluttering flag, the circle of green and beyond, the cleanliness of straight horizons, and the sea.

And felt the vertigo. For here the beach had been uplifted 140 metres by the relentless tectonics. Soft and pale cliffs fell away in front. You could feel the immense drop-off in the air, and could picture too one of the world's most elegantly turned icons, the international golfer, who has struck the little white ball and holds his poise, hitting to the edge of the world.

We walked into the golf shop, seeking permission to drive farm roads to the Cape itself where the terraces were still more pronounced, but were told the farm was private. We struck up conversations later with workers outside the shop, and managed to extract the farm manager's cell phone number, but couldn't raise him. Next day we went in and booked a trip to the Cape with Gannet Safaris, the only way you could now do the trip.

— We call this bank managers' corner, cried Jo Fisher, as she drove her cream van around a corner with drop-offs either side.

— It's an interesting arrangement that people have with their bank managers, so have a look down there. If you can't get rid of them there, then have a look at the other side.

Jo turned out to be a keen tramper and involved with the Mountain Safety Council. Her brother had sold the farm but in the selling had written in access rights for the family. She walked often through the bush and across the gullies. She painted landscapes, and painted well. She had a great sense of elan. Did we want to step out of the vehicle to peer over the edge of the Cape Kidnappers cliffs? We did, and the white van departed the farm track towards that edge while I found my foot, notwithstanding my complete confidence in the driver, quietly pumping an ephemeral brake pedal.

We drove on towards the Cape on a farm track that sloped steeply down. We stopped for shells embedded in the dark cutting alongside. Keith was speculating on the various Pleistocene layers, and Jo, aware by now that she had a geologist aboard, was enjoying the conversation.

— Could that be loess? Now loess is a good word — lurking loess.

A farm worker on a quad manoeuvred past with hair-raising precision, and we wound on down to a stream. Jo's steady commentary continued.

— Now we have a river crossing, a water feature for you.

We followed the farm track around, climbed in low gear to the Cape, and Jo stopped the van beside a low chain fence. The same wide horizon, the same sea and airy updraughts, but the terrace now was brown and studded with shallow depressions, many of them cupping the iconic white birds with the honey-coloured heads. Other sleek adults stood ignoring but occasionally succumbing to the entreaties of their grey fledglings begging for food. An onshore wind gusted the ammoniac smell of fish and Jo was quick to the colony's defence.

— The smell of seven thousand humans together for nine months of the year with inadequate sanitary arrangements would be pretty interesting as well.

Beyond the gannet colony, the Cape stretched away as a sawtoothed ridge descending to the sea and a single tusk-shaped island just beyond.

— Remember Maui, the god, said Jo, pointing to the saw-toothed ridge and island. He was out fishing with his brothers in their canoe, and that's his grandmother's jawbone, or what's left of it. The proper Maori name for Cape Kidnappers is the tooth of Maui or the hook of Maui, Te Matau a Maui. When you're down there in a boat, it's enormous.

I explored down steep slopes to spot another whole plateau of gannets, and when I returned, the van's side door was open, Jo had set out cups of tea and biscuits, and Keith was holding her in thrall with an impromptu discussion of the Cape's geology. He was laying out a map that showed the area in 20-metre contours.

— Good, said Jo. I know about contours.

— The terraces come from sea level changes, said Keith, pointing out the one we were on, and another higher terrace.

He laid out the standard diagrammatic cross section of the downgoing plate.

— Isn't that exciting, said Jo. What a thrill. That's a nice simple diagram.

— It's grossly oversimplified, said Keith. It's just to show you what's happening.

— I'll lie in bed at night and worry about my total idiocy, said Jo.

— This is what we're calling an accretionary prism, continued Keith. All this area is mud scraped off the Pacific Plate. Either think of it as one edge of the old continent, Gondwana, pushing that way, or the Pacific Plate coming this way as a conveyor belt. As it goes down underneath us, the gloop on it gets scraped off.

— Oh, of course, so it's like shavings coming up, said Jo. So every time I file my fingernails I think of accretionary — what was it?

— An accretionary prism, I said, newly returned and keen. You need to embed that in your memory.

— Oh yes, deeply embedded, said Jo. A concrete prism is deeply embedded. I can see that as some kind of wonderful poem.

She shook the tin of biscuits at Keith and me.

— Here. Have another biscuit and come back to reality.

TAUPO
Calderas

In 2000 we completed the first of Te Araroa's Waikato tracks between Meremere and Rangiriri. Sir Edmund Hillary was our patron and it seemed proper to have him open it. It seemed proper too that we should have a local stone, engraved with a local poem. As to the poem, the Topp Twins, Waikato born and bred, volunteered three lines from their song 'Turangawaewae' —

> *A sacred place to be*
> *By the banks of the mighty Waikato*
> *As she flows to the sea*

Then I rang the New Lynn company Trethewey Granite and Marble, explained the need for a stone, and said it had to be local.

— We have a one-tonne boulder of ignimbrite we will donate, said the voice on the phone.

I'd never heard of ignimbrite, but read up on it in case I was called on to speak at the opening. Ignimbrite is a product of rhyolitic volcanoes. It was named by a New Zealander, Patrick Marshall, who noted also its high silica content.

Silica is the most common component in all volcanic rocks, and it's the silica that determines how fluid they are in molten form. The magma chambers of basalt (about 50 per cent silica) are deep and hot, and the magma breaks the surface as a fluid that can leap and flow red, like a display fountain.

The magma chambers of andesite (55–65 per cent silica) sit higher in the crust. The chambers still take in basalt at their base, but as the magma ascends it cools slightly, and minerals crystallise out, raising the silica content of the remaining melt. When andesite magma breaks the surface, its crystal density and its slower lava build high volcanic cones.

The magma chambers of rhyolite (over 70 per cent silica) sit only a few kilometres below the earth's surface. Basalt is entering below at 1250°C, but as the magma moves to the top of the chamber the temperature drops to 850°C or less. Still more crystals form, and separate out from a melt now so viscous that a single crystal might take ten years to sink a single metre. Yet still, time passes, the crystals fall downward into a crystal mush, and the melt purifies. The melt is a nightmare product. At over 70 per cent silica the magma is pliable, but only in the way you might call concrete reinforcing iron pliable. It holds its gases close, it bulges only slowly, it's the Methuselah of all magmas, but when it does erupt, it does so with the force of an asteroid strike.

New Zealanders tend not to know the names of their rhyolitic volcanoes or to acknowledge the source of the pumice that sits on the side of their baths or floats alongside the plastic duck, but they can hardly be blamed for that. The volcanoes are often nothing more than a hole kilometres wide, filled with water and renamed as a lake.

Rhyolite eruptions are rare. In 1912, the Alaskan volcano Novarupta blew out 15 cubic kilometres of magma in just 60 hours, the biggest eruption of the twentieth century. Birds fell from the sky, and darkness, laced with lightning and thunder, came early to the fishing port of Kodiak 170 kilometres away. One woman passed away there in the town's sultry air, but no one else died.

The area was remote and the first three-man party to explore up to the caldera didn't arrive until 1916, climbing up from the south-east and exploring onward to gain an outlook north-west over a wide, long and deep flow of pumice stretching into the distance and still venting its gases high into the air. They named it the Valley of

Ten Thousand Smokes. Later expeditions saw also what they called 'sand flows' — what would later be called 'pumiceous pyroclastic flows' or 'ash flows' or 'welded tuffs' — rock-hard deposits ten or more metres thick.

What then, in the upper reaches of the Waikato River, was a 1930s Public Works geologist to make of whole cliffs of such hardened ash layers? Patrick Marshall had done service as headmaster of Wanganui Collegiate. He saw the cliffs and he reached for his Latin: *ignis* — fire, and *imbris* — storm. He elided the two, and the new word stuck. Ignimbrite was an apocalyptic rock, and in one masterstroke Marshall captured both the howl of its origins and its present geological silence. Most of the central North Island was held in ignimbrite's rigid embrace.

The Austrian geologist Dr Ferdinand von Hochstetter journeyed across the Taupo Volcanic Zone in 1859, and named it. He recognised as volcanic the 'yellow pumice-bearing tuff' that would later be called ignimbrite, and the pumice. He believed the profuse array of rhyolite he encountered had been erupted under the sea, the first stirrings of a volcanic edifice that reached from coast to coast, and through which arose the two mighty andesite pillars of Ruapehu and Tongariro. Hochstetter's descriptions in *Geologie von Neu-Seeland* gave grandeur, scale and international recognition to the immense rhyolitic lodes of the Taupo zone, but he didn't recognise the source: volcanoes so big and violent they'd simply fallen in on themselves and left big round holes — the calderas.

In the New Zealand manner, the subsequent exploration of the rhyolitic volcanic zone was driven by economics, not wonder. In 1872, America had placed Yellowstone's geysers, mudpools and sulphurous rocks under legislative protection as the world's first national park even before there was any recognition of the huge caldera at its heart. In 1887 New Zealand put Ruapehu, Ngauruhoe and Tongariro under national park protection, but the rhyolite centre, with its geysers, mudpools and sulphurous rocks, was left to its own devices. Tuwharetoa and Te Arawa owned most of it and where they didn't, or where Maori sought collaboration, the government set about developing its economic prospects.

DSIR scientists began work on the ignimbrite plateaux that lay to left and right of the calderas to identify the mineral supplements best suited for agriculture, or forestry. The government financed the drill holes that would prove the geothermal fields. Amidst such work, the charting of the calderas proceeded as much by stealth as design. Soil Bureau scientists Alan Pullar and Colin Vucetich worked outside their brief, often in what they'd call 'secret correlation missions' to map and date the ash layers. Geochemists examined and dated sections of the geothermal drill cores. By the 1970s, DSIR scientists had mapped four calderas with fearsome outputs. Two of them, Okataina in the north of the zone and Taupo in the south, showed repeated geologically young or even historical activity, and could be regarded as still active.

The British volcanologist George Walker arrived on a James Cook Fellowship in 1978. Walker had a brow that folded forward like pahoehoe lava over deep-set eyes, and brought with him an extensive knowledge of world volcanology. He was a brilliant field geologist who'd studied Icelandic volcanoes live, and explored the geologically recent craters of the Azores, and the ancient outpourings of lava at the Deccan Traps. He could examine the clast size of fall deposits and estimate plume heights. He could establish isopachs of fall depths on land and compute the fallout lost at sea. He could calculate eruption rates, and speeds of pyroclastic density flows. He could do pure research without the constraint of an economic outcome, and he began to paint a disturbing three-dimensional picture of New Zealand's rhyolitic eruptions.

Walker's first New Zealand paper referenced Vucetich's work

around the Okataina Caldera's Rotoiti eruption and pronounced its estimate of 50 cubic kilometres of magma 'conservative'. The paper would claim the Okataina eruption had produced the greatest airfall deposit, and the highest eruptive column, that the world had seen in 50,000 years. The mushroom cloud was tens of kilometres high, the head of it 200 kilometres wide.

And then he turned to the Taupo caldera. Scientists have dubbed Taupo's biggest eruption Oruanui, 'Big Pit'. Twenty-five thousand five hundred years ago, the eruption despatched 530 cubic kilometres of magma into the sky, taking with it another 420 cubic kilometres of the surrounding rock, excavating the bed of the modern Lake Taupo to a depth of three kilometres. A second eruption 1800 years ago, the Taupo eruption, was smaller in output, some 35 cubic kilometres. This smaller eruption was the focus of Walker's next study. Oruanui's enormous magnitude was not then exactly known, but Walker did know it was big, and he carefully distinguished between Oruanui's magnitude — the total volume erupted over months — and the Taupo eruption's intensity. Taupo had lasted only a few days, but at its explosive climax had erupted more material, faster, than anything Walker had ever analysed before.

Walker's paper was titled 'The Taupo Pumice: product of the most powerful known (ultraplinian) eruption?' The 'ultraplinian' in the title reflected his search for a terminology sufficient to describe Taupo and other New Zealand plumes. Plinian eruptions were named for the plume that rose 30 kilometres above Vesuvius in AD 79 before it collapsed and drove a pyroclastic density current over Pompeii. The plume at Taupo, Walker believed, carried ten times more magma aloft before its column collapsed and drove ash and debris with what he called 'spectacular mobility' across the countryside. To name that effort as 'ultraplinian' was the least you could do.

Assisting Walker, digging down through tephra layers in a

hundred locations, then pausing to mop a high oolitic forehead while Walker peered at the stratigraphy from under his pahoehoe brow, was a young PhD student out of London's Imperial College. Colin Wilson was just beginning a career that would produce many of the most-cited papers on New Zealand volcanology.

Walker would win the McKay Hammer, New Zealand's premier geology award, for his work on ignimbrites in 1982, but by then he was professor of volcanology at the University of Hawaii. Colin Wilson would win the McKay Hammer in 1986 for his work on the Taupo eruption, and go on to become a professor at Auckland University, then Victoria, and to work alongside some of the best American volcanologists.

When I awoke in the morning, I opened the door to grab the morning paper, looked out over the lake, and glanced left to see that the two back wheels of the classic gold Mercury Cougar in the next unit had been clamped.

What idiotic Taupo District Council functionary, asked a sleepy brain, what petty parking tyrant would dare clamp Prometheus, bringer of V8 fire? Then it occurred to me that a chick magnet this powerful would also be a thief magnet, so of course it was the owner who'd clamped the wheels. Come to think of it, volunteered my still awakening brain, maybe Taupo wouldn't want the whole car. Maybe Taupo would just want the mag wheels. Or not. 1970s rock music drifted out of the unit.

Colin Wilson poked his head into the motel unit soon after. He was on his way through Taupo to — in his own words — enrage and frighten a congress of geochemists in Rotorua, but he'd agreed to take the time on the way through for a tour of the Taupo Volcanic Zone. He'd brought with him the stainless steel spade I'd recognise later as his basic tool of trade.

Colin angled the spade into the hatchback, and looked up.

— New Zealand is embarrassed by its geology, he said. It covers up its history with grass.

As the lakeside town's new arterial bypass came up to completion in 2011, Colin had put his case that the road cuttings remain as a testament to Taupo's extraordinary geology, and he'd suggested the Taupo District Council should not grass them over. He'd been invited to take a camera through and record what he wanted, and then the cuttings were hydroseeded, regardless. What petty functionaries, what roading tyrants, dared clamp a green hood over the head of Prometheus, bringer of fire? If the occasion demanded, Colin was known to take to the grassed road cuttings with his spade. He had once been asked by an aggressive contractor what exactly he thought he was doing, and had looked up and replied

— My job.

He cast a sardonic glance at the manacled Cougar as we pulled out.

— Very nice. And the loudest sound you'll hear is the vortex in the petrol tank. We've improved fuel efficiency since then by a factor of three.

He'd just come back from fieldwork in Yellowstone National Park, a landscape that suppurates with geysers, boiling mud and steaming sulphurous vents, the afterglow from the last eruption. Yellowstone sits on a hot spot, a plume from the mantle that has penetrated the North American Plate. The volcano has produced a lot of rhyolite. Its last big explosive eruption, 640,000 years ago, had blown 1000 cubic kilometres of ash across the American Midwest. Its caldera, mapped only in the 1960s, was 120 kilometres long and 60 kilometres wide, about the size of the rhyolitic centre of the Taupo Volcanic Zone.

As we left Taupo township, crossing the Waikato outlet and heading north up State Highway 1, I asked Colin how Yellowstone compared to the Taupo Volcanic Zone.

— There's two yardsticks. One is the magma and molten rock

that stays beneath the surface, and fuels the geothermal systems. Yellowstone is ahead there. If it was a horse-racing thing it would be by a neck, 5.3 gigawatts versus 4.2 gigawatts. The other measure is what comes out of the surface and spews over the landscape. The Taupo figures are still not well understood because a lot of it is buried, but Taupo puts out about two and a half times as much material as Yellowstone over the same time period. The Americans make a lot of fuss about Yellowstone being a supervolcano, and yes it is, but we have bigger and bolder and best. So have a nice day.

Bigger, bolder, and best was going to be around us all day. The plan was to climb Mt Pureora. I'd stood on the Pureora summit in a previous year in a southerly so cold your limbs got slow, but my memories were still vivid. To the south the three volcanoes of the Central Plateau had stood crisp and clear. North lay the smooth volcanic plug of Titiraupenga, and beyond it the white steam flag of the Kinleith mill. East lay Taupo, but that was the last of the simple physiography. To the north-east the land was anonymous and humped, suggestive, uneasy, like pigs asleep under a blanket. That way lay lava domes, ignimbrite plateaux, collapsed caldera walls, cauldron blocks, pumiceous dunes — the whole kit and caboodle of the silicic system, much of it packing heat.

Mt Pureora was on Te Araroa's route, the best of the viewpoints out over the Taupo Volcanic Zone, and I'd been advised that Colin Wilson was the right man to stand on the same summit, to tease out the zone, sum it up.

We drove the Whangamata Road, headed west across the top of Taupo. Kinloch village lay down below us in the bay, a tidy little development of both permanent homes and baches, its golf course designed by Jack Nicklaus. Pumiceous bedding, with its soft contours, well draining and easy to work, was good for golf courses and cemeteries.

The greywacke bedrock was over three kilometres below us. If you brought up a vertical cross section of what lay above the bedrock, there'd be broken rhyolite ramparts of the Whakamaru caldera that had last erupted 350,000 years ago, dumping something over 1500 cubic kilometres of magma onto the landscape, and above that the overlying ash layers and ignimbrite of the two big Taupo eruptions, and uncounted others.

— We're inside the Whakamaru caldera right now, said Colin as we drove on west. It's just too big to see.

We turned up Link Road, driving through the tall rimu of the Pureora Forest Park and the drooping mountain cabbage trees, then stopped before an old road-cut. Without notice, we'd passed out of the Whakamaru caldera and were now standing in front of its product. The road embankment was dimpled with age and weathered to an elephantine grey. Colin swung his Estwing rock pick into its flank and levered off a chunk. He held it in both hands and addressed it with both affection and despair.

— This was once a proud pumice and ash marching across the countryside and incinerating everything in its path, and now look at it.

He smeared the pumice between his fingers.

— It's rotted itself. It's just fluff. If you grab pumice when it's fresh, it's nice squeaky glass with little crystals in it. But dissolved in the magma in the chamber are volatiles, water and carbon dioxide, but often fluorine, chlorine and sulphur. Depending on the composition of the magma, depending on how the thing behaves if you bring the material out, bring it to rest, and it's still hot enough, the glass in the pumice crystallises itself to fine-grained material which kicks out the fluorine, the chlorine, the sulphur, the water, the CO_2. Those are the nasties and so those noxious juices come stewing up through whatever is above it, and they rot the pumice. So this is what you get reduced to. It's clays, it's fine-grained — look at it.

Pureora was by then just three kilometres distant. The mountain was 1.8 million years old, an ancient outrider of the Taupo Volcanic Zone. Like the nearby Titiraupenga it was andesite, with the strength and height to cleave the Whakamaru blasts and leave the deposits of pumice at its foot smoking and seeping, the iron in it blooming to pink, the ignimbrite hardening beneath it to yellow.

I asked Colin how deep he thought the pumice was.

— These are the upper ramparts of something that goes down, in this area, probably about fifty to seventy metres.

We parked the car at the foot of the summit track. The weather looked fine, but big trees all around were stirring in the breeze. A southerly was pushing a cold occluded front along the Hauhungaroa Range.

At 1000 metres we came across pumice lying in banks within the forest.

It was pumice from the Taupo eruption, the same pumice whose 'spectacular mobility' had aroused the admiration of George Walker. Colin picked up a chunk, and turned the vesicular white rock in his hands.

— Just gorgeous. See how it's been stretched out? It's heading into the vent, its frothing up, and just being pulled apart. It is to be praised. It makes the Mount St Helen's blast look like a teddy bears' picnic. This one was probably travelling at sonic velocities for the first forty kilometres and helped knock out twenty thousand square kilometres of countryside in ten minutes.

Our boots resounded on boardwalk and we reached the uncertain realm where the trees shrank. Hallucinogenic white shreds had begun to stream through the trees. Maybe the trees weren't shrinking, maybe you were simply getting bigger. You could project whatever you wanted onto the summit of Pureora, because when we did finally reach the trig, our surroundings were purely white.

There was wind at the summit. There was cloud. We ate lunch hunched in the shelter of stunted vegetation, and we waited, as rapporteurs of the real, a decent half hour for the weather to clear. Colin recalled a previous visit to the summit.

— It was a gloriously fresh day. You could see the Hunua Range, and through binoculars you could see the individual windows in the hospital at Hamilton from the top here, ninety kilometres away. Absolutely glass clear. You could see Taupo, Tarawera. I'm not sure whether I could see the top of Putauaki. You could obviously see Taranaki, Ruapehu, Ngauruhoe. I can't think of any point in this general vicinity where you get as good an overview of the main guts of the field.

— You can point out to those who stand on the summit that the greywacke east of Taupo, the Kaimanawas, goes up to 1.7 kilometres above sea level. Below Taupo they haven't reached bedrock at minus 2.7 kilometres below sea level. So you can give people a very clear idea of the scale of the chasm that's formed below us.

— You can explain the pumice we've seen all the way up is the widespread flow from the 1800-year-old Taupo eruption. You can tell them we know the time of year that eruption occurred, from the fruiting bodies. You can tell them of a third-hand rumour that someone at Waikato University was studying insect fauna at the Pureora Buried Forest and pinpointed the time of day the forest was knocked down. So — based partly on that rumour around insects — we have the shock wave coming through in the late afternoon, around late March or early April AD 232. The only thing missing is to find the copy of the *New Zealand Herald* to know what day of the week it was.

And then, not for the first time on the Te Araroa trail, we agreed to press on, and in the writing of it later to tell people what they might have seen had the weather been more kind.

TAUPO

We dropped down on Highway 32 towards Mangakino and the largest of the calderas, except that it was entirely gone. The output had covered half the North Island. When a storm uncovered a fossil forest embedded in ash at Takapuna Beach in mid-2014, geologists suggested Mangakino as the tephra source. The volcano had erupted something like 2000 cubic kilometres of magma, and 1000 cubic kilometres in one explosive episode that laid ignimbrite from the outskirts of Clevedon and Maraetai in the north and down to Cape Kidnappers in the south.

We stopped to overlook the basin and picked out the vast product of this and other Taupo volcanics. The characteristic level surfaces of ignimbrite marked out the Mamaku Plateau, downwarping from there into depths beyond the reach of either road-cut or field geologist, so there was no beginning to the stratigraphy. The field geologists could reverse-engineer Mangakino's extraordinary radiance and it came back to this shallow basin, but that was all. The caldera walls were gone by erosion or burial, or both. The Mangakino caldera, vast as it was, had been mostly overprinted by the later Whakamaru eruption. In the mid-2000s, Mighty River Power had drilled three kilometres into the Mangakino strata to test hydrothermal potential. The drill went down through 500 metres of sediment from a lake once bigger than Taupo. Then it wound onward through 1.8. kilometres of ignimbrite and still hadn't reached the greywacke bedrock.

We descended into the Mangakino basin, into perhaps the greatest crater in the world, and there wasn't a sign of it, but it made you think. Like a tenebrous god at its obscure work, the nameless shred of land later known as Aotearoa had punctuated prehuman and human history with darkened skies and dropped temperatures. This crater and the others since had lain a chill fingerprint on *Homo erectus* (Mangakino 1,000,000 before the present), and again on *erectus* as they spread from Africa into Asia, and Europe (Whakamaru 350,000 BP). *Erectus* would be extinct, but *Homo neanderthalensis* and *Homo sapiens* would gaze together at the darkened skies of Okataina (61,000 BP), then the Neanderthals would vanish and *Homo sapiens* alone, with its bone needles for sewing, spears for hunting, charcoal for cave art and beads for adornment of the dead, would look up to the haze of Oruanui, or perhaps, in Australia, turn towards the sound of it (25,500 BP), *Homo sapiens* moving right along through its stone age and bronze age to agriculture, settlements, finally to the Chinese and Roman empires, which were overshadowed, however briefly, by Taupo (1800 BP).

And right then a precise human history was still unfolding in its glory as a white Toyota Corolla cruised on a tarmac highway across the old Mangakino crater. Cattle stood in the fields, one of their number licking an ignimbrite outcrop. Sheep huddled against a bank for shade. There was ryegrass to feed them, and seven-wire fences to hold them in. Pine plantations covered the hills. Pylons swung electric energy away from the two hydroelectric stations. A gas station was pumping petrol. A general store stood on the corner of Highways 30 and 32, and someone suggested coffee.

We went in and the plenitude of *Homo sapiens* closed around us. The general store had all the essentials you might expect. The Huntly coal and the fire starters. The eggs, the sliced bread, the baked beans and groceries of all kinds, but beyond that too the ornaments of a fervent and triumphant species. A profusion of plastic ferns, lace stoles, a red jug, a choice of Mexican sombreros, slicka pads, tiger-skin-patterned throw mats, Chinese lanterns . . . and nor did the variety stop with the millinery and the merchandise. It was there in the range of food. The store had more pastry wrapped around more types of meat than you'd ever expect in a pie warmer, a wondrous array of ice cream flavours, an Xbox game firing up on a large screen in the corner, and as we sat down finally at the Formica and chromium pipe tables to address our coffees, someone came by toting a bright pink marshmallow square.

— That, said Colin, looks positively evil. They should perhaps make it the same colour as the 1080.

Add firearms to the list of *Homo sapiens'* achievements, and add to its list of deeds, murder. As we crossed the Waikato on the dam top and drove up Highway 30 on the true right of the Waikato, we were travelling the same road a farmer trundled on his tractor in March 2013 when a blue Cherokee Jeep started following along behind. The farmer continued up Highway 32 towards State Highway 1, expecting the Jeep to pass, but the vehicle hung back and turned off the highway finally onto Tram Road. It stopped at a roadworks. The bridge across the Waikato here was under the control of a stop–go man, George Taiaroa. The driver of the Jeep shot him dead, and accelerated away over the bridge.

Time and weather had erased one of the world's biggest volcanoes, but the Mangakino basin was still full of relict violence. Across the river, in the town's gravel pit, was a boulder as big as a house. It wasn't a local rock, and nor could there be, in these low latitudes, any question of transport by glacier. As we drove, you could see more big rocks, lodged in the roadside grasses, or half-lit within the colonnaded darkness of ascending pine trunks.

The walls of a river gorge rose in sheer leaps around us, 300 metres high either side of the road.

— As best we can judge, said Colin, this gorge has been the outlet for the Taupo Zone for longer than three hundred and fifty thousand years. It seems to have been a low point, and so every time Taupo sneezes there's a big flood goes through here.

The Oruanui eruption had blocked Taupo's outlet to the north, and the lake behind had begun to swell. Colin had traced a wave-cut shoreline in the cliffs above Taupo, around 140 metres above the lake's present level, a vast weight of water that overhung the Mangakino basin. For a few years to decades that was the status quo, with a new river outlet draining west off the high-standing lake. And then by earthquake, or intolerable pressure, the ignimbrite dam at the old northern outlet broke.

— The Cake Tin in Wellington, said Colin, as we drove on through the gorge, holds about one million cubic metres. The Waikato River at its present flow rate would fill it in about eighty minutes, so it's a useful comparison. The twenty-five-thousand-year flood out of Taupo would have filled the Cake Tin in about three seconds. That was the flow rate.

We passed Tram Road, still following the river through, and ahead of us then was the first sight of Pohaturoa. The 540-metre lava dome had breasted the 25,500 BP flood. That tumultuous slurry had stripped away all but the central volcanic pipe, which rose almost sheer from beside the river in 100-metre leaps. We came up to State Highway 1. Sunk into the road margin alongside was a stone big enough to crown the entire top of Pohaturoa.

We stood at its foot, and an intermittent stream of traffic blatted past.

— This doesn't belong here, said Colin, patting the monolith, and only about one-third of it is above ground. It's a very massive rock. It's got gold in it. It's hydrothermally altered. It came from the Ohakuri Dam area, about six kilometres upstream.

A behemoth, rolled by water. I climbed the great boulder and was saluted at the top by the air horn of a passing semi. Later I wished I hadn't been so keen, and remembered the air horn as less salute, more lament. When Carter Holt Harvey cleaned the pines off the sides and summit of Pohaturoa in 2000 they laid bare the remnant dirt floors of whare. Pohaturoa was a near-perfect defensive pa and had served as a sanctuary for Ngati Kahupungapunga, the people of the pumice lands. But Ngati Kahupungapunga would never lay claim to its old territory. In one of its detailed studies to assign ownership, the Waitangi Tribunal found Ngati Kahupungapunga had been routed off Pohaturoa in the sixteenth century by Ngati

Wairangi coming through from Kawhia, and, in the words of the tribunal, 'exterminated as a tribe'. The Tribunal heard evidence requoted from Hare Reweti Te Kume addressing the Maori Land Court in 1868. 'The great stones (at Ongaroto) on which the bodies were cooked can be seen still.'

Fifteen kilometres east of Taupo on State Highway 5, we stood before a road-cut that contained the complete stratigraphy of the 1800 BP Taupo eruption. The 35-cubic-kilometre output made it the third largest eruption of the last 5000 years, but it was the most violent, and on a blue afternoon, Colin Wilson's flashing spade rendered its layers pristine.

— The Romans are doing their thing, said Colin, walking himself back in time. The Chinese are doing their thing. My ancestors are painted with woad and hiding out in bogs.

He picks at the base of the stratigraphy. Here is the layer he found in his early fieldwork, and named the Initial Ash. It's pale, one centimetre thick, innocuous, a first plume distributed mainly to the east by a prevailing wind, where it lightly dusts the Kaingaroa Plateau. The next eruptive column surges skyward from an island or promontory in the lake and its signature ashfall is dry. Then wet eruptions roar from vents along a 10-kilometre axis down the eastern side of the lake. One layer comes down so wet it forms its own rivulets and small gullies. A hole in the road-cut is a long-gone branch that has gathered, as does any tree in a good snowfall, a half-round of white ash on the branch. That small white hump is preserved in this wall like an icon. Then a layer of small round lumps singly or glued together, puttering down like hail.

The eruption then starts to accelerate, the vent switching back to dry again, and an eruption column perhaps over 50 kilometres high rears over the central North Island, scattering coarse pumice further than any other historic eruption on earth.

This is the plume that first caused George Walker to reach for his 'ultraplinian' superlative.

— It's going along, said Colin. It's an impressive eruption, and then something very nasty happens.

The second sequence begins. The eruption switches into a discharge of such scale and intensity that geologists shy away from its physical description, and tend to lend it instead a mathematical alibi of output per second. The emission rate is vast. Arising now from the earth is a grey mothership tens of kilometres long and kilometres deep. It's attended by scintillating outriders of lightning, and the heat given off by the particulate body of it assures its steady rise. It becomes a great ascending bolt of earth and magma, and the shock wave that radiates out from it knocks down the forest at Pureora. In Colin Wilson's envisaging, however, neither that rapidly expanding ring of shocked air, nor even the sparkling bolt at its centre, is the nastiness.

The eruptive column climbs into the stratosphere. It holds about 30 cubic kilometres of material that's erupted in perhaps six minutes. The strength of the column is its heat, and as that dissipates the column begins to collapse. Far below, the airfall deposits of past hours have draped the countryside white, smothering all vegetation and wildlife, but preserving the topography, and now, with the collapse of the column, the nastiness begins. The pyroclastic blast that scythes out from the base of the column levels the topography as it goes, expanding swiftly, beheading the shrouded hills, jumping the valleys, slicing outwards at near-sonic speeds, leaving behind a thin veneer that later somewhat dumbstruck investigators will call low-aspect-ratio ignimbrite. Over an 80-kilometre radius, all fin, fur and feathers will perish.

To the south of the blast is a mountain rampart, but that does not contain it. The sonic speed and weight of it drives the expanding saucer of debris south, 1300 metres upslope past the Emerald Lakes on Tongariro, and beyond to the 1600-metre mark on Ruapehu, or higher. In the west it surmounts the summit of Titiraupenga and rushes through the Mangaokewa and Waipa valleys before its seething momentum dies. In the east it rushes through the plains and river valleys of what will become the route of the Taupo–Napier Road, surges beyond the Titiokura Summit and slides to a stop on the long downhill to Hawke's Bay. The northern edge of the same sheet makes a final upward leap through the Hemo Gorge and falls quiet finally on the outskirts of Rotorua.

Yellowstone frightens America. There's a lake bed that's gradually rising, two volcanic domes that show signs of inflation, a free-ranging bison herd, and bears. As the park bison migrate down the highway to summer pastures, a popular panic may spread along that same highway that the animals know something more than the SUV drivers, and they may do three-point turns and step on the gas, out of there. Yellowstone documentaries carry foreboding music, and their CGIs show pyroclastic density flows that roll forward at supersonic speeds and envelop the viewer. The last big caldera eruption was 640,000 years ago, and the Americans now have sufficient data from Yellowstone's previous large caldera-forming eruptions to venture a return rate of between 600,000 and 800,000 years, and to warn of a pattern of smaller eruptions.

It's not much of a guide, but New Zealand's rhyolitic eruptions are even less predictable. There's no single chamber to erupt and slowly recharge, but six that qualify as still active. The chambers sit within a rift zone that is pulling apart at the rate of 10 millimetres a year. The rifting is tectonic, not volcanic, but volcanologists now think the rifting bears upon the eruptions, making them unpredictable as to return times, essentially chaotic. And the heat that's driving the rhyolite chambers around Taupo is a factor of 10 higher than other volcanic arcs around the world.

The first suggestion that New Zealand volcanism might be

intimately tied to its restless tectonics came with the analysis of the Tarawera eruption of 10 June 1886. The basalt eruption was modest by rhyolite standards, but it killed over 100, mainly Maori villagers close to the mountain, felled by ballistic bombardment or enveloped by the base surge as Lake Rotomahana erupted. Further afield, at Te Wairoa, the same base surge was felt as a fierce wind that swayed chimneys and caused flames to blurt straight out from the wood-burners, but 14 died later that morning at Wairoa anyway, most killed by structure collapse and burial under the weight of an ash and mud fall that lasted six hours.

The eruption attracted its own myths. A mysterious canoe seen in daylight a week before the eruption by a tourist group heading across Lake Tarawera was named, even before the eruption occurred, the phantom canoe. That story was embellished later. The Auckland Art Gallery holds Kennett Watkins' oil painting that shows the ghastly craft and its benighted crew piercing a boil-up of orange steam and clouds on the lake, heading for a dark mountain where a rising moon domes the summit. And then there was the Te Wairoa tohunga, Tohutu, who twice warned of destruction on the way.

Such were the superstitious stories of popular record, but when it came to the science, the head of the New Zealand Geological Survey, Dr James Hector, didn't do a lot better. Hector had departed Wellington aboard the *Hinemoa* the day of the eruption, steaming through a yellow fog in the Bay of Plenty on his way to survey the damage. He spent two days exploring around the volcano, and his report concluded none of the disturbances that preceded the eruption were useful for predicting it, and that Tarawera was a 'purely hydro-thermal phenomenon, but on a gigantic scale' — a steam explosion.

He was badly wrong, but part of his report retains its resonance today. The eruption rifted across three separate summits, then ran on south for 10 kilometres, blowing out Lake Rotomahana and, beyond it, the row of fumaroles named since as Black Crater, Fairy Crater, Inferno Crater, Echo Crater and Frying Pan Lake. Hector speculated the eruption might be a local fracture triggered by movement 'along the great fault lines that traverse the North and South islands in a northeasterly direction'.

In 1981, a paper by Ian Nairn and Jim Cole noted that the basalt intrusions into the rift on Tarawera were consistently stepped to the left. The two geologists described the phenomena as *en echelon* formations consistent with dextral shear on an underlying masterfault. Simply put, if you imagined the two sides of a rift as the long stringers of a ladder, and the basalt intrusions as its rungs, then any sideways slippage of the right-hand stringer would twist all the rungs at a consistent angle. The rift and the basalt dikes were apparently interacting.

In 2010 Colin Wilson was part of a team that brought chemical analysis to bear on the huge eruption from Mangakino, 1,000,000 BP. The research was predicated on the fact that magma in separate chambers may have different compositions, however subtle. The team's analysis of the wetfall deposit from the Mangakino eruption showed, at different stages of the same eruption, the systematic tapping of three independent magma chambers.

— It's magma coming into the crust versus the ripping apart imposed by the tectonic plate structure, said Colin. They all talk to each other and we're trying to understand the language.

The Oruanui eruption of 25,500 BP also, though perhaps briefly, tapped a second magma chamber. It was one of the biggest of the big white monsters, but it was not continuous. The stratigraphy shows 10 different fall units, some no more than hours apart, but others whose textured surface indicates a hardening over days or weeks, and one surface, near the start of the eruption, that shows the reworking of the surface by wind and water, possibly too disruption by burrowing insects. The break may have been months or even years.

— How do you do that? asked Colin Wilson. How do you take five hundred cubic kilometres of magma, gas saturated, ready to roll, and turn it on and off, like the beer tap in a pub?

The speculative answer was that here too magma chambers connected one with another as the tectonic rifts opened and closed. Analysis of some pumice in early fall deposits from the Oruanui eruption picked up traces of the dark mineral biotite. The Oruanui magma chamber contained no biotite. Fifteen kilometres northeast of Oruanui sits another magma chamber entirely. It erupted around 27,500 BP, not a big eruption, but sufficient to identify its biotite fingerprint. The implication is cautiously stated, as is proper in science, but is clear enough: in 25,500 BP an earthquake opened a passageway between the chambers, and the north-easterly magma trundled through to join the Oruanui eruption. Two magmas, previously unrelated, and probably quiescent, got to meet each other and started to party.

The centre of the North Island has been jittery for longer than people now remember. The crust is thin here. The tectonic plate out to the east pulls sideways, and beneath it the magma shrugs and stretches, and there's plenty of movement underfoot.

In 1895 an earthquake collapsed most of Taupo's chimneys, a wave estimated at 0.6 metres swept across the lake, landslides blocked roads, and springs at Hatepe halfway down the eastern side of the lake spouted fine pumice. Many residents and visitors camped outside in fear of the damage. Whether the event was caldera unrest or a regional earthquake is not known.

From April 1922 to January 1923, Taupo endured an earthquake swarm, peaking at the rate of 57 shakes a day. Parts of the shoreline just east of the township sank over three metres, water fountained from cracks in the ground, chimneys collapsed and the Taupo town clock stopped.

Another swarm shook the town from September 1964, peaking in December, with the epicentre moving out towards Horomatangi Reef in January 1965 with a uplift in that vicinity tentatively measured at nine centimetres.

Another swarm came through in 1983, and the western part of the lake shore around Kinloch rose 5.5 centimetres.

— No one took much notice of it, says Colin. What are you going to tell them? Kinloch has gone up fifty-five millimetres. What can you do?

GNS Science has six permanent seismographs, a strong motion site at the Taupo Police Station, eight telemetered GPS units, and shoreline sites that use the lake itself as a spirit level to detect tilt. Colin Wilson sees limited security in the system.

The odds against a big rhyolite eruption — 10–15 cubic kilometres out of the Taupo Volcanic Zone — are 5000 to one in any one year, but the odds against any rhyolite eruption, big or small, are 900 to one, a vexatious fact for Civil Defence.

GNS volcanologists can trigger an ascending scale of Volcanic Alert Levels. VAL Levels 1 and 2 cover volcanic unrest, including changes to hot springs, earthquakes, steam eruptions, escaping volcanic gases, uplift and subsidence. Then the VAL scale tips into volcanic eruption alerts. VAL 3 is a 'Minor', VAL 4 a 'Moderate' and VAL 5 a 'Major' eruption, with an implicit responsibility to evacuate the town.

But exactly when do you evacuate? Colin Wilson puzzles about that. He puzzles about what might have ensued had there been a town called Taupo extant in the year AD 232, early March. The Taupo field geologists of that era paint themselves with woad, but Colin accepts those blue faces as an eccentricity of the time, and grants them anyway the same geological knowledge he and his colleagues have collected, up to but not including AD 232. He walks the town back to the beginning of autumn that year, and it's obvious something is afoot, but what?

Is there a substantial earthquake swarm? Yes, but the swarms come along every 20 years or so. Are there new steam vents, or a change in hot pool activity? Yes, but nor is that unusual. Has the land around Taupo risen? Yes, a little, but that too has happened many times before.

The district geologist settles for VAL 1, minor unrest, but then the Initial Ash erupts. Now what? The assembled blue faces of the field geologists can assure their boss that since the big Oruanui blasts of 25,500 BP they've identified some 26 separate eruptions out of Taupo, and 18 of those are altogether larger and more worrisome than the somewhat innocuous plume now ascending out the window. How much of a fuss should the district volcanologist make about it?

'Okay, it's a definite eruption, but let's make it a VAL 3, a "Minor Volcanic Eruption".'

Anything more than that, and the ur-New Zealand dollar is likely to drop by a third and the ur-NZX40 to shed 50 per cent of its value.

— If you were relying on the older records, pre-AD 232, there's no way you would have predicted what happened in the Taupo eruption, said Colin. It changed character very suddenly. It was unique. And that's the problem. How would you know?

Then there's another factor — the safe return of the people. When does an eruption finish? Ten years or so after the deadly sting in the tail of the Taupo eruption, after the ur-Waikato dams have been cleaned out, when the ur-turbines again begin to spin, when blue faces again sip coffee in the streets of a rebuilding town, what does the district volcanologist have to say about the lava dome that begins to push up from the lake bed? For that, too, is part of the Taupo history. The pumice on the flanks of the rising dome peels off in lumps as big as boats, and they pop steaming to the surface of the lake and are nudged ashore by the prevailing westerly.

The lava dome is still there. Sediment has raised the lake floor many metres, and only the top of the dome now peeks through — the Horomatangi Reef, gushing boiling water. The great lumps of pumice are still there too, stranded now in the fields alongside State Highway 1 as it exits south from the town. Colin Wilson stood beside them, laid on hands and said

— So here's another eruption cycle maybe ten years later. What advice do you give someone? These are the sort of details that are too nasty for economists and government people to deal with, and it's only by the mercy of eighteen hundred years that we missed having to deal with it.

— Whenever scientists talk about giving warnings about this or that, have at the back of your mind, what are they warning against? Think of Christchurch. The Darfield earthquake happened and the generalised warnings went out that the biggest aftershock will be about one magnitude smaller than the main strike, magnitude 6.1. No one said, 'Ha! The bad news is, it's going to be distributed onto another fault that will monster the city centre with the highest peak ground accelerations recorded in virtually any historical quake known.' This is what's missing from the whole hazards and warnings thing. A real appreciation of the cussedness of Mother Nature.

We stood on the Mapara highway on the eastern side of Taupo, above Acacia Bay, and embedded in the mossy embankment overhead, too high for cleaning down with a spade, was a rock the size of a small car. There was no impact crater under the rock. It didn't go up and come down. During the Taupo eruption it came out in a flat trajectory from a vent 13 kilometres distant to lodge here along with a rabble of other large rocks and stones.

— Part of the beast, the monster, the big one, Colin Wilson was saying. To carry blocks that size this distance. Very few other eruptions have managed that. Mount Saint Helens didn't manage it. Just colossal momentum . . .

To be able to see, in its own time, the rising of the huge column and the gaseous slurry that delivered the rock was to be a dead man watching. The only safe vantage point from which to watch the Taupo eruption was space.

A black SUV sped down the highway towards us, out of another world.

— Step in a little bit, said Colin. I don't wish to lose any clients.

We felt the rush of air, and then he stared after the vanishing vehicle.

— Bloody Taupo, said Colin. This beautiful volcanic site, and they're just blissfully unaware. It's money, trout and tourism, and, 'Yeah. It's a volcano — so what?'

Malcolm Haitana was a big man in a small house. The sign painted onto the outside of his small lakeside bach read *Fawlty Towers* and he stood now braced either side of the ranch sliders, shaking the structure like a dog with a rat, throwing his weight right and left, up and down, so the whole place shook, the ornaments rattled, and my chair swayed. Malcolm Haitana was creating a serious disturbance, and he kept at it.

— This is what it was like Geoff. I was asleep. The ground was rumbling. I thought it was a train coming. Then this mighty SHUUUSSSSH. I thought it was a jet plane. The rumble, the shaking, the noise was extreme.

He'd jumped out of bed, rushed out those same ranch sliders, tripped over a line of shoes on the porch and rolled out onto the lawn. It was hard on midnight, 6 August 2012. He looked up at the mountain and saw the ascending column, a black outline against the stars. A big white flash went racing up the column, then globes of white-hot gas ascending. Red-hot rocks arced out either side. He raced back inside. His partner, Linda Graham, was standing there, and they held each other.

— The fear never hit you. I was just uncertain. Is it going to get worse?

Malcolm and Linda were as close as anyone got to the hydrothermal explosion out of the upper Te Maari crater on Mt Tongariro, four kilometres away. It began with a debris avalanche as the crater heaved, and tonnes of material rumbled away downhill. The Te Maari crater had been dormant for over 100 years, and it needed to clear its throat. The vertical eruption plume shot into the sky, and the whole of that first and most violent eruption lasted just 16 seconds.

For Malcolm and Linda there was none of the cool assessment that ties a natural event down to its numbers, just something close to frenzy. What do you do? What do you take? Nothing. Just get out, and do not wait around in the confines of a blind shotgun shack for the roof to fall in. They took off, running towards the camp entrance, two of them together and the big unfriendly mountain. Linda wanted most just to see another human being. She wanted reassurance, but the camp is seasonal. There was no one. The mountain boomed, they stood at the entrance and then came the single thought, thin as a straw in a sea of doubt and darkness — *Our passports!* That thought held them, somewhere on the edge of flight. They waited 20 minutes. By some trick of the wind that night there was no ashfall down at the camp, and nor had they come to any harm. They went back to the bach, but still on the edge of flight, passports to hand, and ready to run. Another half-hour went by and the mountain was calming. A posse of locals came by to check on them. They went to bed.

We jumped into Malcolm's runabout, and trolled for trout out on Lake Rotoaira, looking back at the huge flat bulk of Tongariro, the whole top of it largely gone from past eruptions, and the gouts of steam from Te Maari trailing away in white threads across its flank. Ketetahi Hut stood just over a kilometre west of the crater. The eruption had ruined the hut. Te Maari's boulders had smashed through the roof, through the bunks, and through the floor. A plastic water tank outside the hut was broken into curving green shards, and the whole of the wreckage then covered with a dirty ash.

Te Araroa's route piggybacks upon the Tongariro Alpine Crossing across the Central Plateau, and I'd walked it three times. Of the three active volcanoes on the Central Plateau, Tongariro had seemed to me a somnolent mountain, sunk into old age, her great head slumped forward on her chest, and as if to compensate for the ravages of great age, I'd once written, she was hung with a kind of jewellery — the blue lake with its ice-encrusted surround, the emerald-green lakes, and the flat round cameos of old craters.

Do not patronise old ladies. Seventy thousand trampers a year used the Tongariro Crossing and the mountain was now proven well capable of tithing those passers-by. I looked to see what the geologists had made of the Te Maari eruption, and the most compelling study was by Rebecca Fitzgerald, from Canterbury University.

Her team plotted from the aerial photographs 3587 impact craters between 30 centimetres and 10 metres in diameter. The team went onto the ground later and mapped a further spatter of projectiles, smaller but whose craters bespoke too their capacities to kill and maim. Te Maari had thrown out over 13,000 serious missiles, and the team mapped the trajectories and the impacts, denoting craters over 6 metres in diameter with large dark circles, and in three categories grading down to lighter and smaller. They marked up the hut, and its bombardment was obvious. The mapping showed too how thoroughly the crater had mortared the track itself. No walkers died. The missiles fell around midnight and by simple chance no one was staying at Ketetahi Hut. It seemed to me both the Tongariro Alpine Crossing and Te Araroa had simply ridden their luck through New Zealand's most recent, and rather tiny, eruption.

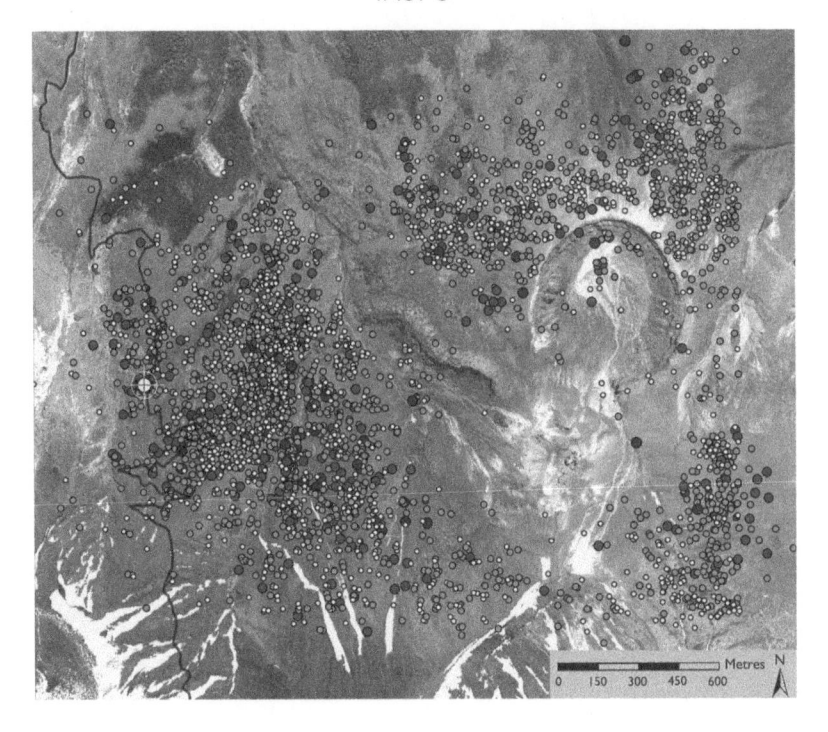

WELLINGTON
Earthquakes and fossils

I bought my copy of *Geomorphology: An Introduction to the Study of Landforms* from the local second-hand store. It was once a popular book, and the price was still holding up — $35 for the 1958, revised seventh edition. Its author, Charles Cotton, had been a professor of geology at Victoria University from 1921 through to 1953.

New Zealanders once bought Cotton's book by the thousand, and it was also an international textbook. Keith Lewis, for one, had read it as a course book while still a student at Reading University. The book sourced most of its wide range of landform examples from New Zealand and I liked that — *Mount, with due attention guys and gals, your own doorstep, and you have entry to the world* — while conceding too that Cotton's time had passed. The vocabulary was dense and difficult, and I found the book too polite. It moved through a surface world that was neither geology with tectonics below nor geography with people above.

The book laid out the stages of landscape evolution, largely a record of steadily increasing erosion. A landscape could be infantile, youthful, sub-mature, mature, post-mature, or senile, but that seemed to lead into a trap. Beyond senescence there could be no such thing as a dead landscape, so the only way forward was to start another cycle. Rejuvenation, no less.

Glancing through *Geomorphology* I found a black and white photograph of Wellington looking across the harbour to show the steep fault scarp behind the wharves and the motorway. The scarp that strikes SW–NE across the city, rising abruptly within one kilometre of the central business district, and 600 metres from the New Zealand House of Representatives and the Beehive —

the Wellington Fault. The book instanced the fault's signs of geologically recent 'rejuvenation'. Oh, happy hanging ravines, for here was a sign of an old landscape become young again. Ah, sweet shutter ridges, for those abruptly truncated forms also signalled the youthful renewal. Any 1950s quake on the Wellington Fault, before earthquake standards rose, would have wrecked New Zealand's capital city. Rejuvenation, I thought, didn't seem quite the right word.

I looked up one of Cotton's later papers that dealt more specifically with tectonic force. He proposed an original Wellington K surface, planed flat by the weather or the sea. You could see the K surface everywhere. It was Wellington's original simplicity, still obvious on the ridges and plateaux above Makara where the Meridian wind farm now stands, or at Tabletop north of Porirua, and disguised elsewhere by the erosion Cotton was so good at describing — a deeply dissected 'post-mature', even 'senescent' landscape. But it wasn't just erosion. Furrows ran through the K surface on SW–NE alignments, sufficiently long and straight, wrote Cotton, that pilots used them to navigate. The longest and straightest of them was the Wellington Fault.

Cotton found evidence of a 60-metre offset on the Wellington Fault. He described the sideways movement as 'transcurrent drift' and noted also that vertical uplift along the western side of the fault had formed a formidable scarp, and that the downthrown eastern side had formed Wellington Harbour. He was right, but he also believed the K surface was a peneplain, eroded flat by weather, and that may have caused him to underestimate the size of Wellington's uplift. Later research would suggest the K surface had been planed flat by the sea, and that both simplified and increased the calculations of uplift.

Cotton got the overall shape of Wellington tectonic movement close to correct. The Wellington region, he wrote, had undergone a 'paroxysm of deformation'. Yet whether by the demands of careful scholarship or the innate restraint of a somewhat shy geologist, Cotton's language remained subdued. There was that somewhat wafty 'transcurrent drift', and he noted 'There are signs of yet another, though as yet milder, spasm of continuing activity.'

The book *Geomorphology* and Cotton himself had shaped the tenor of his time, no question. The painter Colin McCahon was somewhat outside the tenor of his time, but also used *Geomorphology* to ramp up his own landscapes. They didn't seem much alike, the professor, blind in one eye from a teenage accident, with his careful pen and ink, and the bloodshot artist and his bold oils. McCahon would overlay Maori words onto his landscape. Cotton would eradicate at least one Maori word from his. In the 1940s, a number of geologists were circling around the concept of Wellington's ancient flat surface, and one at least had begun to use the name Kaukau as a referent. Excluding the Tararua and the Ruahine ranges, it was, after all, the highest local summit where you could trace the alleged peneplain. Cotton wrote that the name he proposed, the K Surface, would both recognise but also remove the 'uneuphonious' Kaukau from the lists.

Cotton's pen and ink drawings gained the truths he wanted by denuding the landscape of its vegetation and buildings. He exposed the surface by subtraction. McCahon gained his truths by abstraction. His big steep forms exposed the surface then reached through it to the slop of fluids underground, and the unseen faults that slashed the land.

— He got the darkness, said John Begg as we drove into the Hutt Valley.

With John at the wheel, we continued slowly along the Hutt Road watching the black suburban sidestreets of Petone, then Alicetown, stretch away at right angles, straight and flat, except that as each one departed the Hutt Road, it dipped away down a short slope. We were driving parallel to the Wellington Fault. The short slope was the fault itself, and we noted commercial buildings, built right onto that small slope, that would, in their end time, rise perhaps one metre and wrench sideways five, spraying out bricks like an overripe seed pod. Later we walked beside the Hutt River. Here, undisguised by the tarseal of suburbia, the smashed rock of the Wellington Fault rose to the surface in long strips of riverine breccia, as rough to the touch as crocodile skin.

We climbed to the Kaukau summit. The wind was flattening the brown grass south. It was soughing in the telco mast behind us. We went down into the lee, found a spot out of the wind, and settled in. Wellington lay at our feet.

— Beautiful old town.

— Yeah. Beautiful.

— The clouds parting just sufficiently that we can see the city down there.

— Yep.

The green clasp of Wellington's town belt. Its efficient transport hubs, trains moving in and out through the tunnels in the hills, crooked pathways connecting the steep levels of it. Interisland ferries moving out from the harbour into Cook Strait. The steeper its landscapes, the more the wind tugged at its hair, the more the city seemed to embrace a civil order neat as its own commutes, and the enclosing circle of its bright stadium.

We leaned back on our elbows amidst the grasses of Mt Kaukau, and John Begg was not trending towards civil order. He'd been growing out his beard, so he claimed, to mark the upcoming 175th birthday anniversary of the famously hirsute and self-educated government geologist Alexander McKay, one of New Zealand geology's folk heroes. John's idea was that all his colleagues who could grow beards should grow beards, then gather round the McKay portrait on the day, smile to camera, and there you'd have it

— the unreformed face of New Zealand geology. John's Newtown house was large and rambling, and the geologist padded around in it, swerving occasionally to grab a beer for his guest, his bare feet imprinted with the clear untanned V of the customary jandals those same colleagues called his Cook Island work boots. A large Samoan Church across the road sang in praise of the Lord every Sunday, and there was something of a Polynesian generosity of spirit within the Begg house, with adult children coming and going, and his keenly political wife Justine fighting her corner with aplomb. John was presently refining a 3D model for Christchurch that would show householders there the geological structure beneath their street, and the engineering they might need to make it safe, but back in the 1990s he'd taken on the job of completing the first 1:50,000 geological map of Wellington.

He propped himself on an elbow, and pointed across to Port Nicholson's eastern entrance.

— That's Baring Head. You can see a few trees on the tip. That's more or less where the lighthouse is and that sits on the top of a marine terrace. So that's about sixty-five metres elevation.

I looked, and recalled one of Charles Cotton's sketches. Cotton had stripped away the trees and the lighthouse, but 100 years ago he'd sketched just that marine terrace. What Cotton didn't have was the means to convert such geological markers to an accurate tectonic movement over time. John Begg did have the means.

— That marine terrace was the highstand shoreline of 125,000 years ago. We found a shoreline of the same age in drill holes at Petone where it lies a hundred metres below sea level. That gives a difference of a hundred and sixty-five metres between the elevation of this shoreline at Baring Head, at the harbour mouth, and Petone at the base of the harbour.

In short, within just 125,000 years the shoreline that stretched 15 kilometres from the top of Wellington harbour to the bottom had twisted significantly, risen in the south-east and sunk to the north-west.

The K surface yielded a deeper measure of Wellington's torque. John Begg and others had established that the K surface was something over a million years old. The Kaukau summit was part of the K surface and stood at 445 metres. That same surface lay under the sediments on the harbour floor, some 600 metres under the sea.

— A good guess for how far the K surface has been displaced by the Wellington Fault in the last million years, said John, is one kilometre of vertical offset and a lateral displacement of between five and twenty kilometres.

Wellington shakes often and, disturbingly, in response to faults that strike directly towards, or parallel to, the city. New Zealand Company settlers got their first big shake in 1848, a magnitude 7.2 temblor, but the epicentre was the Awatere Valley, some 100 kilometres distant. The Wairarapa Fault ruptured in 1855, a magnitude 8.2 or above, and in 1942 a magnitude 7.2 centred near Masterton was followed five weeks later by a magnitude 6.8. Those earthquakes all damaged the city, but the epicentres were all distant, and neither they, nor the magnitude 6.5 Cook Strait quake of 2013, were top of mind for John Begg when he said

— The taniwha is there, but it doesn't show its face very often. Once every thousand years or something like that, so one generation in twenty has to pick up the pieces. Civilisation as we know it goes back five thousand years and in that time you've had perhaps five ruptures of the Wellington Fault.

We ate some of last night's curry leftovers with our fingers and paid tribute to the Kiwi ingenuity that emerged from Wellington's rolling quakes. The lead and rubber bearings of the New Zealand-invented base isolators, also the Seismic Design Code for engineers, which listed the Z factor. That factor varied according to region, responding to criteria like distance from active faults, and earthquake return times on the faults. Each different Z number called into play its own complex equations, even down to the so-called 'Spectral Shape' factor, which provided separate mathematical values for hard rock as a building foundation, through to deep soft soil, and all points in between. Engineers were bound by the Z numbers to tailor seismic design to the differing levels of seismic hazard.

The Z factor for Wellington, Porirua and the Hutt Valley was high, but not as high as Otira, or Arthur's Pass, or the townships along the Alpine Fault, and a few North Island towns. Lay that Z ranking alongside the number of people affected by an earthquake, however, and the economic fallout from it, and Wellington led the country's rankings for risk exposure.

We lay propped on our elbows, and praised the citizens below. They lived with the risk and their engineers mitigated the risk by rerouting electricity, gas and water supply lines, even dispersing water supply into small self-sealing suburban reservoirs, for the citizens' guerrilla survival. In Christchurch I'd seen the disturbing sight of pigeons flying freely in and out of the cathedral's wrecked nave. Wellington would also get its quake, and would there be owls in the Beehive?

I'd taken a previous interest in Wellington's faults. I knew them by name. Te Araroa followed the Ohariu Fault between Porirua and Mt Kaukau, then the Wellington Fault into the city. As the cloud began to block the view from Kaukau, as the streaming cloud began to scrub out the mast behind us and wind it up to a low moan, I was pleased we'd talked through the Wellington Fault, its million-year age, its last rupture between 400 and 700 years ago, its return times of around 1000 years, and the fact that the earthquake it generated was likely be in the magnitude seven range rather than an eight. I said

— So the Wellington Fault is the one the city should fear?

— It's one of the ones Wellington should fear.

That was when John Begg came up with the name of a fault I'd never heard of, the Plate Interface Fault, and I made the mistake of thinking he was talking about the subduction zone, 110 kilometres south-east.

Not at all, said John. The Plate Interface Fault was just 25 kilometres away. Twenty-five kilometres, that is, straight down, and not just closer than you ever imagined, it was also stuck.

— Locked, said John. If it was slipping you'd get a whole lot of micro earthquakes, but our instruments show nothing is creeping. That's a good indication that the surface may rupture in one of the great earthquakes at some point. Something over magnitude eight.

His emphasis was on the 'great'. The PIF threw a whole lot of calculations out the window. The geologists didn't know when the last PIF earthquake was, or even if there'd been one. They didn't know when the PIF might come or how it would come. It might activate the Ohariu or Wellington faults that had ploughed their million-year furrows through Wellington's K surface, and equally it might do something quite else. Really, they didn't know what kind of a beast it was.

— We're sitting on the lower end of a locked interface, said John. It's sobering stuff, but the whole of Japan is on a similar scene. Chile is sitting on the same thing. If you're looking for the big story of Wellington, that's it. The subduction interface and the way it transfers that strain to the surface.

We tramped off the Kaukau summit, and finally came on down Tinakori Hill. This was Te Araroa's entrance into the city, a descent down the face of the Wellington Fault scarp, and we stopped just before Elephant Rock, overlooking what had been, for decades, the official residence of the New Zealand prime minister.

Grey, two-storied, a sweeping drive, a canopy built out over the drive, where even in the worst of weathers a dignitary might alight dry from the BMW and hurry up the ascending staircase. We could see a square shadow on one of the roofs where a chimney had been closed off. Another two chimneys either end of the big house had been condensed to squat forms from their elegant originals. When Wellington's minor quakes roll through, the chandeliers inside this house tinkle, and I'd put in an Official Information Act request: had the house ever been earthquake strengthened? The reply was, 'Premier House does not come under the same requirements as commercial buildings to be strengthened and has not been strengthened.'

— The Wellington Fault runs through on this side of the house, said John, pointing straight down. It'll rupture along a plane about ten centimetres wide, but there'll be deformation for ten or twenty metres each side of that. If there's more than one G of vertical acceleration, then anything sitting on this hillside will be thrown up into the air, and when it lands back on shaking ground, it's likely to roll. The crush zone, I'd say, would come up to the back door, and undoubtedly all these houses down there will also get more extreme acceleration and more severe ground shaking. You'd expect more deformation here than, say, across on Mount Victoria.

Mt Victoria! I lifted my sights. The New Zealand flag was streaming south-east from the Beehive 500 metres away. Mt Victoria was another 2.5 kilometres beyond that. I'd been concentrating so hard on the near-field effects it just hadn't occurred to me that as far as the eye could see, the whole city was going to shake.

By agreement, I arrived at the museum early, before the public was admitted. Dan Ruck the exhibitions manager took me to the first floor. The lead conservator Robert Clendon and the collections manager Carolyn McGill were waiting. The technician Brad Welch approached the glass case with a key and someone said

— Is the alarm off?

Brad checked. Okay. He opened the case, and Robert moved in. He snapped on rubber gloves and took hold of the mount, one blue hand fingers and thumb extended, coming down as a gentle grab from above, one hand a gentle support from below. It was a practised professional movement, and together the blue hands drew out the mounting box along a long stainless steel rod. I leaned forward to watch and felt a hand on my shoulder.

— Stand well back.

Protocols. A clear space behind the handlers to forestall any bump or other impedance during the carriage of the white mounting box as it was backed out of its glass case, delivered to the long table set up nearby, and set down there with reverent care.

The heavyweights of the Te Papa exhibitions and collections team fanned out from the table, and I moved up. Robert came alongside me and gave a permission which had been unclear to that point. I could touch the tooth.

To get this far had taken a long time. I'd emailed Te Papa. I'd asked to touch the tooth, but what seemed to me like a simple request proved difficult. The tooth was currently on display, not presently stored, and for security and insurance reasons the request was declined. I girded myself for the fight, and one of the big guns to fire was the man who'd co-authored the definitive account of the tooth's history in a 1997 paper, 'The status of Gideon Mantell's "first" Iguanodon tooth'. Aside from his mathematical talents, Garry Tee is also New Zealand's greatest gumshoe explorer of nineteenth-century science, and of New Zealand's place in it. The tooth had come under his scrupulous gaze when it was part of the

Museum of New Zealand's geology exhibition in 1971. 'How nice,' he thought. The story as he understood it then was that Mantell's wife, Mary Ann, had found the tooth, and he thought, 'I never expected to see Mrs Mantell's tooth in New Zealand.'

Seven years later he went back and it was gone. Gone where? Disappeared off display and into the vast vault of the Museum of New Zealand, and upon enquiry it seemed no one knew not only where it was, but also what it was. Gone into the maw, undistinguished, unloved. The tooth that was the first clear evidence of the Age of the Dinosaurs was in peril, uncatalogued in the dim and crowded spaces where objects may lose their identity because of staff changes, and years later go out with the rubbish. It was a scandal, and Garry insisted it be found. The vaults were turned upside down in the hunt. Weeks later came the report from the director — 'We've found it!'

As the man responsible for the rediscovery of the tooth in 1978, wrote Garry, drawing himself to his full height in a note intended for the acting CEO of Te Papa, he believed Mr Chapple's writing would be enriched if it could include a report of handling 'that immensely significant fossil'.

That note, and a few other persuasive points: I came to Wellington on the understanding that the mission might be possible, but only if the necessary team of supervisors could be gathered together in time. As it turned out, they could not. I looked at the tooth in its glass case, but couldn't touch. I returned to Auckland. More weeks went by before word came. They were ready.

Someone hung a pair of blue gloves in front of me. Protocol. The salt from human sweat would erode a tooth. I pulled them on, and I addressed the still air with a speech.

— I pay tribute to Nicolas Steno and James Hutton, for between them they brought the idea of paleontology and stratigraphy to geology, and deep time. Then the great fossil hunter Gideon Mantell, and the anatomist Baron Georges Cuvier, and the geologist Sir Charles Lyell. Those three all handled the tooth. To me, the tooth is a junction with some of the greatest figures in world geology.

Maybe the speech was a kind of apology. I hadn't set out deliberately to banjax the Te Papa protocols, but I seemed to be bumping into them, left and right. As we'd begun that morning's endeavour they'd said no to any recording of conversation. It was in the employment contracts, and needed a specific media clearance. Protocol.

Or maybe the speech was simply pompous. The guardians looked back at me, impassive. I leaned forward. There sat the Holy Molar. It was big. It was brown. It was 135 million years old. The long blue finger approached like something guided from afar to choose a landing site on a strange planet. Hill or valley? The molar's grooved occlusal surface gave me a choice and I wasn't being granted enough air to do both. I chose the valley. I touched the tooth.

The contemporary images of Nicolas Steno show a man with a Roman nose, whose locks tumble down to his shoulders. Born the son of a goldsmith in 1638, Steno lived a short and brilliant life and could easily have been lost to geology, for from age 19 when he entered medical school in Copenhagen he proved such an able student and so dextrous with his dissections that he was guaranteed a public audience, acclaim, and usually a job, in whatever European city he chose. He was a practitioner who could dissect and hold out the secret of a horse's eye, a dog's leg, and, as it turned out, an otherwise undiscovered salivary duct within the human cheek, but he was also an observer.

As a student he'd visited the cabinets of curiosity that were the forerunners of the modern museum, and a feature of most European courts. Stuffed polar bear cubs slung from ceilings might rub shoulders with a dodo, the carapace of a tortoise might hang from the wall beside a stuffed armadillo. The cabinets sometimes displayed fossils chipped out of rock, and the Danish Royal Kunstkammer at Copenhagen had one distinctive type, the *glossopetrae* — literally 'tongue stones'. You could argue whether these small white triangular shapes were natural growths within the rock, or the jagged edges of lightning bolts, or were fallen from the moon, but plenty of them had been collected around Malta, so you could also invoke the biblical story of St Paul's shipwreck at Malta, the serpent bite he suffered there, and the saint's revenge that turned the tongues of Maltese serpents to stone.

'At the castle,' wrote Steno in the cool observational style of his early journal, 'I saw the famous room which is decorated with manifold conches and shells of various animals. There were four doors with these emblematic pictures: Death, Doomsday, Heaven and Hell. There were various beautiful sights, etc.'

He moved from Copenhagen to Amsterdam, then to Paris, and over the Alps. Like many natural observers before him, he saw and wondered at the shells entombed within alpine rocks. As the prevailing explanation had it, they were deposited during Noah's Flood, but Leonardo da Vinci, who never tired of measuring anything that moved, had already noted that cockles move between six and eight feet a day. The 40 days and 40 nights of the biblical deluge, da Vinci suggested, was hardly enough time for a cockle to plough its furrow the 250 miles inland from the Adriatic Sea to Monferrato, where fossil cockles had been found.

In Italy Steno fell in with the Medici brothers. He curated the brothers' cabinet of curiosity in Florence, but his primary role was as a physician, researcher and dissector. In that role he performed the public dissection of a great white shark, caught off the Italian coast.

The shark's teeth resembled the *glossopetrae* Steno had seen in cabinets of curiosity as far back as Copenhagen. In Copenhagen, too, he'd watched experiments where sediments precipitate out of water. It seemed to Steno a straightforward observation that for a hard object — a tooth — to be enclosed by another hard object — a rock — then the one that retained its original shape must pre-date the enclosing medium.

The tooth came first, the enclosing rock second, and that meant the hard rock must at one time have been softer than the tooth. That suggested a fluid sedimentary influx, and the idea was radiant with consequences. If rock was formed by precipitating sediment, then each layer would be horizontal when first laid down. Steno didn't openly challenge a church wisdom that counted back the generations of the Old Testament, and decreed a creation date for the earth of around 4000 BC, but even a glance at the ascending layers in any tall cliff, each layer a different sedimentary pulse, would have opened the clerical claim to serious doubt. Even a glance at rock layers tilted away from the horizontal would have suggested an intriguing duration.

What Steno was firm about seems, looking back, obvious and even risible, but in a world that believed maybe rocks just grew it was a breakthrough for observation as a tool more useful than dogma. As the rock pile ascended layer by layer, Steno claimed, it got younger. That principle, later called superposition, became basic to geology's most powerful early tool — stratigraphy.

'Pity the arse,' wrote Hutton in 1774 to an Edinburgh friend on one of his long rides along the bad roads of Scotland, England and Wales, 'that's clagged to a head that would hunt stones.'

A qualified doctor, gentleman farmer, and active member of the Scottish enlightenment, Hutton believed that the world was slowly laid low by erosion, and was then recycled and consolidated by compression and upthrust by heat. The erosion he observed on his own farm and the amount of uplift he deduced from the crags around Edinburgh suggested such cycles might be slow, and he set off on horseback to plumb the theory, riding through a theological world that was 6000 years old, or thereabouts, and that still believed Noah's great flood of 4000 years past explained all fossils.

Hutton's letters to his friends were devoid of capitals or punctuation, a litany of humour, sexual innuendo, excitement for his inventive friends like James Watt, who was just then perfecting his steam engine, and over and above all of it his pursuit of the rock samples he called, with dissembling charm and without the usual capital or the apostrophe, 'bibles all wrote by Gods own finger'.

Over 15 years he would study rock strata, delving into what he called 'the annals of a former world'. Amongst the most visually convincing of his former worlds, Hutton found a river-cut bank at Jedburgh, Scotland.

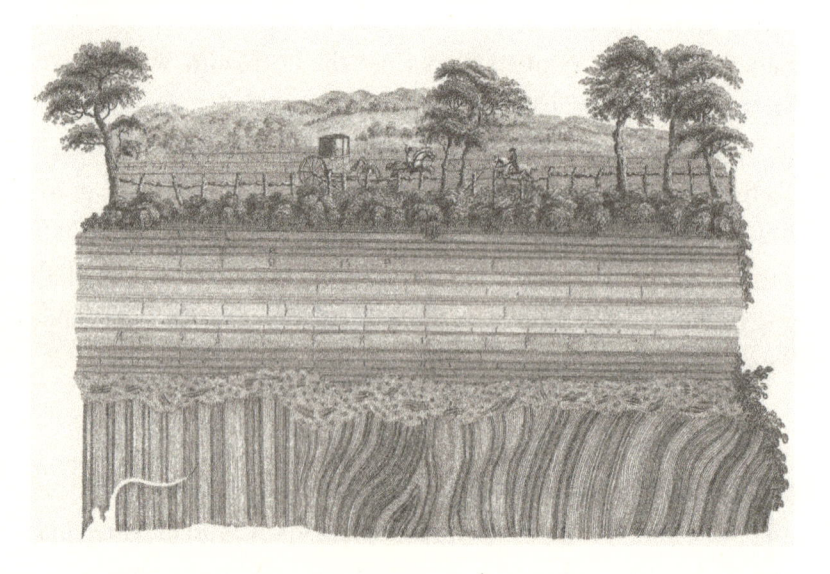

His companion John Clerk recorded the Jedburgh strata in a famous sketch that showed a base of near-vertical greywacke and the overlying layers of horizontal sandstone. The vertical beds suggested once-mighty mountains brought low by erosion, the horizontal ones a later waterborne wash of sand and mud, compressed into rock and now uplifted from the sea. Clerk penned a delicate horse-drawn phaeton progressing along the sunlit veneer of these two lost worlds — the present and prancing third world, oblivious to the consigned darkness of the former worlds below.

Reporting on his years of work in the *Transactions of the Royal Society of Edinburgh* of 1788, Hutton would invoke the far fetch of time implied by mountain building, erosion, subsequent relayering, and continuing uplift out of the sea. Even one such long cycle, he supposed, might be only one amongst a number, each one when judged on the erosion of the present world slow beyond any human history. The world rehearsed its geology again and again, but there never was a final performance, and Hutton ended the paper with a famous last sentence: 'The result, therefore, of this physical enquiry is, that we find no vestige of a beginning, no prospect of an end.'

On 4 April 1796 at the age of 26, the French anatomist Georges Cuvier delivered his inaugural lecture to the National Institute of Science and Arts in Paris — *Mémoires sur les espèces d'éléphants vivants et fossiles*. Elephants were known to Europe at the time, and most of Cuvier's science-literate audience would have already read sufficiently to know of differences in temperament. The elephants of Asia were mild-natured and easily harnessed to such human enterprise as hauling logs, and the African elephant was a much wilder beast. Still, an elephant was an elephant, just as a dog was a dog. Not so, claimed Cuvier. On close examination of the teeth and skulls of each, he could confirm the Asian elephant was as different from its African counterpart as 'the horse from the ass or the goat from the sheep'. They were two different species.

Cuvier then turned to more ancient relics. French armies in America in the 1730s had sent back to Paris large bones retrieved from a swamp in Ohio, and teeth the size of bricks. Cuvier had studied this cache, also the large bones and a tusk the size of an alpenhorn found in Siberia, and he named to his French audience that day two more elephant species: the mammoth, and the 'Ohio Valley beast', the prehistoric animal he'd later name more precisely as a mastodon.

Europe's curiosity cabinets held plenty of fossils — ammonites as big as wheels, trilobites and the solid bullet-shaped belemnites. They were odd, but the prevailing view was that their living counterparts might be found alive in some unexplored corner of land or sea. Cuvier begged to differ, and in his lecture that day took up an idea suggested previously by others, but never in such a prestigious forum nor with such a showman's flair by such a rising star of science. The idea was new, and it was shocking. The remains, said Cuvier, confirmed two more species of elephant, but with one terrific distinction. They were *disparu* — extinct.

One surmises a sharp collective intake of French breath as Cuvier, having spirited his huge and forgotten beasts into the room, vanished them again forever with the click of a finger. But he wasn't yet finished, and he brought the elephant lecture to its climax with a suggestion of violence beyond imagining —

'All of these facts, consistent among themselves, and not opposed by any report,' said Cuvier, 'seem to me to prove the existence of a world previous to ours, destroyed by some kind of catastrophe.'

Cuvier did not intend drifting within Hutton's fathomless time, nor did he accept the Scotsman's somewhat precipitate conclusion that the fossils that turned up in cabinets of curiosity and private collections were, with a few exceptions, more or less a duplicate of existing life forms. By the time Cuvier began work on the stratigraphic pile it had already been divided into three main divisions — primary, secondary and tertiary. Large skeletal fragments were being found within the secondary and tertiary layers, and Cuvier set out to gather the fragments and, by his anatomical skills, to interpret, even to reassemble, the broken forms and to repopulate the ancient rock surfaces with the beasts of their day. He invited fossil collectors worldwide to send him their discoveries, and join him to — as he put it with Gallic flair — 'burst the limits of time' to picture life on earth much as the historian did who recreated the human ages of stone, bronze or iron by the fragments thrown up by those ages. He invited the collectors to go far beyond that modest human span, to hunt in the ancient rock and become with him 'a new species of antiquarian'.

The record is unclear as to whether Gideon Mantell or his wife Mary Ann found the tooth at Cuckfield quarry, Sussex. Mantell was a surgeon in his nearby home village of Lewes, also a member of the Geological Society with a keen interest in fossils. He knew Cuvier had visited Oxford Museum in 1818, examined a lower jaw with one large curved carnivore's tooth still inset that was found in secondary limestones, and pronounced it the jaw of a very large reptile. Mantell's new find came from secondary layers also, which suggested a reptile, but the tooth was different. The flat crown indicated the beast had ground its food. That suggested an herbivorous reptile, but Mantell was cautious. In his 1821 self-published book *Fossils of the South Downs* he said simply that the tooth, and others he'd retrieved since, were 'of a very singular character and differ from any previously known'.

In those same early years, the young barrister and gentleman geologist Charles Lyell visited Mantell's house to view his collection, and the two became friends. The Geological Society's best anatomists had dismissed Mantell's find as the tooth of a large mammal, washed into the quarry rubble from younger strata. Mantell was gloomy about that and Lyell had no firm opinion but was about to visit Paris to improve his French. He agreed to take the tooth to Cuvier.

On 28 June 1823 Lyell duly showed up at one of Cuvier's regular Saturday *salons*. He was then aged 25 and Cuvier was 53.

Cuvier took the tooth from the young Englishman. The French master was already aware of reptile fossils in the secondary rock formations, but he was certain only of reptiles that swam, and others that flew. He knew of the ichthyosaur and plesiosaur, discovered by others, and both marine. He'd personally identified the remains found in chalk layers at Maastricht as reptilian, and would later confirm it as marine, and name it a mosasaur. He'd examined an engraving sent in from Bavaria on strange bones there, and named it as 'ptero-dactyle', a flying reptile. His 10-year-old hunch of an ancient world order of reptiles was being gradually proven. They were in the sea, they were in the air, but as yet he had no good proof of reptiles on land. The Oxford Museum jaw was suggestive, not definitive. Evidence of an herbivorous reptile, by definition a land-dweller, would be the proof. A dominant order of reptiles, land, sea and air, to confirm his hunch. Onward from there, it might confirm another hunch, for what could have ended such a dominance except a catastrophe?

Cuvier took the tooth, and a lot hung on his judgement. He studied it, gave his verdict, and missed his best chance to pronounce, then and there, an age of reptiles, land, sea and air.

hovered over the tooth. It was mounted on an annotated card, and I read the inscription, in Mantell's handwriting —

Left lower abraded molar / This was the first tooth / of the Iguanodon sent to / Baron Cuvier, who pronounced / it to be the Incisor of Rhinocerous.

A small mystery still surrounds what Lyell told Mantell upon his return from Paris to report the verdict, for on the reverse side of the mounting card is a note by Lyell, written much later, in 1859, which responds directly to Mantell's somewhat recriminatory comment on Cuvier's pronouncement and carries on from it —

This however was at an / evening party. The next morning he told me that / he was satisfied it was / something quite different / Sir C. L. 4 Feb 59.

It seems likely that Lyell didn't report those second thoughts to Mantell at the time, but he did report the rhino verdict, and Mantell took it hard but persisted with his explorations in the quarry. He found more teeth and sent them to Cuvier a year later. This time the Frenchman came up trumps. Yes, the teeth indeed belonged to a giant herbivorous reptile, and he would include Mantell's find in his 1825 edition of *Recherches sur les ossements fossiles de quadrupèdes*. That book was the world authority on fossils. Mantell was vindicated and duly celebrated by his Geological Society colleagues, and now he needed to name his find. He visited the anatomical collections of London and was alerted to an iguana skull. The teeth of that dominantly herbivorous tropical lizard were 20 times smaller, but otherwise a good match, and Mantell now had a name for his giant grazing reptile — Iguanodon.

Cuvier died in 1832, and another 10 years would pass before the sharp-elbowed Victorian paleontologist and anatomist Richard

Owen would propose a distinct name for a distinct order of giant land lizards. Owen based the new classification on three English finds: the carnivorous Megalosaurus named by the English geologist William Buckland, the herbivorous Iguanodon, and the armoured Hylaeosaurus, both discovered and named by Mantell. The name Owen proposed was Dinosauria, from the Greek *deinos,* meaning terrible or wondrous, and *sauros*, lizard.

So to the tooth. I pulled off the gloves, and a hand reached out to take and dispose of them.

— No thanks. I'd like to keep these.

They had, after all, touched history, the first direct evidence of the greatest species ever to walk the earth. You couldn't blame the guardians for treating it seriously. I turned to Dan.

— The only people who aren't here are Security.

— They're here. He pointed to twin cameras in the ceiling. There'll be three of them watching this right now.

I'd come to Wellington to look at the tooth, but also to look at the geology. I'd traversed along the Wellington Fault with John Begg, and he'd mentioned the Wairarapa Fault a few times. On 23 January 1855, a rupture along that fault had shaken New Zealand end to end. It was the country's biggest earthquake since colonial settlement and strangely the tooth had a role to play here also, for the best historical account of the quake came from Sir Charles Lyell and one of Lyell's best quake witnesses was Gideon Mantell's son Walter, by then also the inheritor of the tooth, and the man who would later gift it to a New Zealand collection.

As the story went, Gideon Mantell had kept on with his fossil hunting, but at a cost to his career and his family. He transferred his medical practice to Birmingham in the late 1830s but it didn't prosper. He converted the family house into a museum to help finances, and as the fossil display advanced, so the liveable area retreated. Mary Ann, the talented illustrator of his fossils, walked out. Walter left the family, too. He dumped his medical training and shipped out to New Zealand.

Walter became a keen amateur naturalist, and also Commissioner of Crown Lands for the southern district of the South Island. He reconciled with his father, sending back moa eggshells, and bones, and Gideon returned the favour with rock samples to help his identifications and a 50-drawer cabinet to house Walter's own specimens. Walter never quite adopted the settler culture of New Zealand. By the mid-1850s he was sufficiently disturbed at the New Zealand Government's failure to provide the schools, hospitals and reserves promised to Ngai Tahu during the South Island land purchases that he returned to London to settle the matter with the Secretary of State for the Colonies.

He arrived in 1856, and in subsequent months failed to get an audience with the colonial secretary, let alone redress. He did renew the family friendship with Lyell. Gideon was by then four years dead, but Walter knew Lyell of old, and before he shipped back to New Zealand, he would take the tooth, mounted on its card, and ask the world's greatest geologist to inscribe it. The pressing topic during his first visit to Lyell's Harley Street residence, though, was the giant New Zealand earthquake.

Lyell sought out every reliable quake witness passing through London that year. From Walter he gained the geology of the upthrust ranges and the plain below. From Edward Roberts, now returned to London as a member of the Royal Engineers but in 1855 the Clerk of Works for Wellington, he gained a measure of the uplift. From Frederick Weld, a Marlborough sheep station owner, recent MP for Wairau, fine watercolourist and a future New Zealand premier, he drew the South Island effects. The sum of those reports made plain that whole regions of New Zealand had been uplifted and whole regions downpulled.

Lyell published his final account of the quake in the tenth

edition of *The Principles of Geology*, published in 1868, with the following introduction —

> *In no country, perhaps, where the English language is spoken, have earthquakes, or to speak more correctly, the subterranean causes to which such movements are due, been so active in producing changes of geological interest as in New Zealand.*

The Wairarapa rift, he wrote, was 90 miles long. Its shocks were felt by ships 150 miles out from the coast, and the shake affected an area of land and sea 'three times as large as the British Isles'. Around Wellington a tract of land 'not much inferior to Yorkshire in dimensions' had been permanently uplifted between one and nine feet while the South Island's Wairau Valley and its immediate coast had sunk five feet. Anyone seeking the Wairau's fresh water now had to journey a further three miles upstream.

Thus the great quake of 1855: yet looking back, the odd thing was that none of those contemporary observers recorded the quake's most unusual feature. I went to see Rodney Grapes, who'd written a book on the quake, and asked him why.

— No one knew such a thing could happen, he said. And so they didn't see it.

The upheaval itself had witnesses on land and at sea. In Cook Strait thousands of dead fish popped to the surface. On Wairarapa farmland, one witness near Greytown saw the jaws of the earth open twice and shut again with a 'fearful snap'. But even after the tsunami drained back off the Rongotai isthmus, after the dust settled in the Wairarapa, no one saw it. Maybe the farmers were too busy winching their tumbled cattle out from the open rifts to see. Maybe Roberts, who had the training to see and measure, was just too thunderstruck. His pre-quake job had been to survey and complete an all-tide coastal route to the Wairarapa, and returning after the quake he found the task accomplished in a moment by a new beach 30 metres wide. Like everyone else he saw the uplift, nothing more.

And besides, as the aftershocks died away, the ambitious new town of Wellington conspired in its councils and reportage not to make any undue fuss. It was Wellington's second quake since settlement, and no one wanted to frighten off new immigrants. Casualties were light — at most nine people died. The low death toll, the determination not to frighten people, the absence of any specialist geologist, and New Zealand's fast revegetation banished the quake to memory.

The Wairarapa Fault line waited patiently 100 years before geologists began to see the full power of it. Aerial photographs sounded the alert in the 1950s. They showed a wide band of abandoned channels and faulted terraces beside the Waiohine River. The aerials suggested that Wairarapa earthquakes had moved the land not just up, but sideways by spectacular distances. In 1955 Harold Wellman went out to check it on the ground. He had his rough and ready methods, tying cloth to fence posts, checking the lines with a Dumpy level. He reported the land around the river had moved sideways over time, 120 metres. In 1962, Gary Orbell examined other big watercourses and claimed a 212-metre shift. The time span for the shift was large — perhaps 35,000 years — and what the geologists could not easily find was persuasive evidence of sideways movement from a single quake. The big rivers were useful for tracking macro-movement, but their periodic floods smoothed away evidence of any smaller movement, and the question remained: what was the single shift — the strike-slip offset — from the 1855 earthquake?

In 1988 Harold Wellman and Rodney Grapes set off to map the fault again. During that expedition, Grapes was driving north towards Featherston on Western Lake Road. Over 100 metres away but running nicely parallel to the road was the old green scarp of the 1855 quake. Grapes shot the odd glance sideways, and as he approached Pigeon Bush, he saw something. The light was just right and he spotted two cupped shadows at the foot of the scarp.

I drove out along the Hutt motorway. A 2004 study on the Wairarapa Fault by Tim Little and the American David Rodgers remeasured the rift that opened in 1855, and reported an uplift far greater than anything Roberts had seen. Parts of the Rimutaka Range jolted skyward six metres, and that great uplifted slab sloped down from there to Eastbourne, uplifted two metres, and over to the Wellington shoreline, uplifted 1.5 metres. This was the same shoreline, highwalled to my left by the Wellington Fault scarp, that by John Begg's account was sinking. When I put that to him later he'd say what was uplifted here by the Wairarapa Fault would be thrown down again by the local Wellington Fault, but whatever, for the moment the Wairarapa Fault was sufficiently on top of that tectonic interplay for me to drive along the sealed surfaces of its uplifted 1.5-metre harbourside terrace. Thank you.

The Wairarapa earthquake's huge regional displacements led Little and Rodgers to suggest the quake may have been more powerful even than the accepted magnitude 8.2. It was perhaps an 8.4, almost as big as land-based earthquakes ever get.

I drove through Upper Hutt and on over the Rimutakas, looking up at the greywacke cuttings that rose high above the road. Jolt by jolt a succession of Wairarapa quakes had raised this range over 1400 metres in something under a million years. At Featherston, I turned right along Western Lake Road, and the old scarp kept pace until I came to Pigeon Bush. I found the spot. There were two fences to cross, and I turned into the nearest house to get permission. A man came to the door — yes, people came to look at the Pigeon Bush geology from time to time, and were sufficiently a nuisance that he charged $10 for access.

— You really are asking ten dollars for me go across the paddock for a look?

— That's right.

You can't argue the rights of private property holders. I paid the $10, and for my money went and stood on a green knoll between two old stream beds. I didn't have anything to add to the expertise that had crowded onto this bit of turf, but I tried to imagine the thump of it.

The two old stream beds emerged abruptly at the foot of the scarp as if the scarp itself had once produced an upwelling spring at the head of each channel.

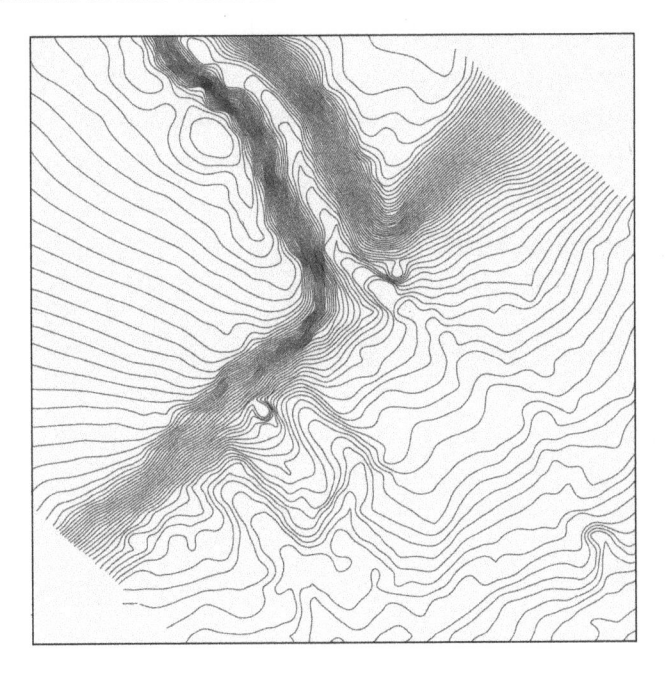

Yet there never was such a spring. To the right of the two beds ran their original source — a quiet stream lively enough to have eroded down into the scarp, and flowing onward through green pasture.

It was flowing as I watched — nothing much more than an open tap might produce — drawing water from a small catchment in the hills behind, but even in spate was not lively enough to obliterate the two beds that lay to its left.

The explanation for the two dry beds was obvious enough. The bed most distant from the present-day stream had been — in geology's language — beheaded by a strike-slip jolt around 1500 years ago. The parent stream had then cut itself a new bed that persisted until the earthquake of 23 January 1855 struck and it, too, was beheaded.

During their 2004 study, Little and Rodgers came to Pigeon Bush and remeasured the strike-slip of those two earthquakes. They used GPS and laser-ranging techniques to produce a micro-topographic map, and measurements that were as precise as the new technologies allow. Around 1500 years ago, the first stream bed had jumped sideways 14 metres and dried. Then on 23 January 1855 the second stream bed jumped sideways 18.7 metres. That was, by a nose, the greatest single strike-slip offset ever recorded on the face of the earth.

MARLBOROUGH SOUNDS AND THE RED HILLS
Ophiolites

The white path leading away from Ship Cove in the Marlborough Sounds is Te Araroa's start in the South Island, and any walker on that track will know intuitively they're walking across a sunken landscape. What were once hilltops are now islands, what were once ridges are peninsulas, and without reference to any geology, the walker on this three-day hike along the thin finger that separates Queen Charlotte and Kenepuru sounds may still resurrect from such skeletal remnants and hold in their mind's eye the Sounds' origin as opulent river valleys.

Such is the physiography of the Sounds, but since it lies above the downgoing Pacific Plate, the obvious question for geology is why this inundated landscape is not, like its compatriot territories across Cook Strait, jostling upwards.

One paper out of 1968 declared that it was indeed rising, that the Sounds were marine canyons, an uplifted block out of Cook Strait. Later researchers disagreed, but retained the idea of a great block, which might not be rising but was certainly being tipped about. The tipping of it might explain some of the queer morphology around the Sounds. The short rivers at the base of the Sounds seemed to lack the capacity to gouge out the depth and width of the great valleys that had existed before inundation. And then there were the wind gaps — long empty corridors through the hills that suggested a large river system had once drained the 2000 square kilometre catchments of the Sounds. The Pelorus River was one part of it, and the whole system had debouched south into the Wairau River near Blenheim. The Pelorus River still flowed, but upended now as the ancient south-aligned landscape tilted north, and opened up to invasion by the sea.

MARLBOROUGH SOUNDS AND THE RED HILLS

The wind gap and the tilt theorists had their critics, and even the majority that came to believe in a northward tilt disagreed amongst themselves. The block was sinking north-west, said one paper. No, it was sinking north-east, said another, and perhaps it was inevitable that Sounds geology was ambiguous. Fault systems that provide an easy land-based measure of movement both up and down were missing here. The Sounds were bounded both west and north by the faults that lie under Cook Strait, and the GNS regional survey of Wellington called that fault system 'poorly understood'. Sea-based evidence was a crock, and the usual land-based evidence for tectonic movement — the marine terraces and other geological markers so useful for measuring uplift — were, within a descending region, progressively destroyed.

In 2010 a radio-carbon study of tiny sea-creature fossils within mud layers reported an average descent of 80 centimetres every 1000 years. Two years later, GPS measurements from GNS stations in the Sounds confirmed D'Urville Island was sinking, but at Tory Channel and Okoha the land was, surprisingly, rising. The GNS team that reported the result cited margins of error, and if not that, then the possibility that a stickiness within the sliding plates might have caused a temporary bulge. No one wanted to contradict strong geological evidence of sinking, but nonetheless the figures were ambiguous.

The Pacific Plate slides under Wellington just 25 kilometres below the city streets. The Sounds lie due west of Wellington, and as the plate slides on below them it's 40 kilometres down and beginning to drop steeply. But this is also the region where the Pacific Plate's trench system encounters a substantial land mass, and jams. Land can't be subducted, and here the sudden accumulation of land along the Pacific–Australian plate boundary forces a transition from subduction to a more elastic sidle of one plate past another, accommodated within the South Island's strike-slip faults.

The shape of that transition is unknown beyond the tremors

of earthquakes, most of them small, recorded here week on week by the seismographs, triangulated as to depth, calibrated as to magnitude. That earthquake record shows not just the quiver of the plate's descent, but also perhaps the jitter of some rogue piece of it migrating south. But it's only a perhaps. The record, plotted as it is with red and yellow dots by the thousand, is finally no more than a blindstick tapping to discern a shape within the twisting underworld below Marlborough, and its best descriptor may not lie with the science at all but with the Latin summary, *Id est quod est* — it is what it is. Within the torque of the transition, a great block of land, slowly sinking.

I finished my table study of Sounds geology, bearing in mind their loveliness from two previous tramps but finally admitting I couldn't put a coherent geology together, and finally I went back to the oldest story of all, which suited the Sounds' darker character of strong tidal laminations through narrow passages, of winds amplified by the long sea channels and turned by the hills into sudden squalls that could overturn yachts, or pull up a waterspout. I went back to the squid.

Two rock eyeballs protrude from the sea at the approach to the Tory Channel, known to Maori as Nga Whatu-kaiponu, the eyeballs that won't let go. In the centuries of Maori precedence in this country, waka crews paddling between Aotearoa's two main islands protected themselves against that gaze, bagging the heads of figurative carvings front and back of the waka, and blindsiding their own faces with karaka leaves, lest the stony gaze enchant the waka and hold it unmoving on the water.

The eyeballs belong to a giant squid, a wheke who once oversaw the devastation of the fishing grounds around Rarotonga and in the time before ours was chased across the ocean by Kupe and slain here, his eyeballs plucked out and the corpse further shut down by incantations that made any resurrection impossible.

The depths beneath the Sounds sag, there are flexures up and down, left and right, and a thousand earthquakes that record the twisting of a ductile body. All these effects are true, and finally we might have no better belief than this: that the form of the sunken sea creature held down by Kupe's enchantment has begun to stir. That the lid begins to lift. The wheke lives!

Later I overflew the Sounds, en route to what's commonly called sunny Nelson. The captain's voice came over the intercom.

— The flight before us just tried to land into Nelson and aborted, but we'll give it a crack.

He gave it a crack. The plane jolted in the air as its wheels jack-knifed down into thick cloud. Bobbling rivulets ran sideways across the window. I strained through cloud to see the Nelson runway.

Mike Johnston was down there, waiting to begin a tramp along the Red Hills. I'd chased Mike two or three times during my year of geology, and he'd been either busy at home or away on some geological journey overseas. He was the man responsible for much of Nelson's geological mapping. He was also the New Zealand authority on Ferdinand von Hochstetter. Mike Johnston actually looked like Hochstetter.

I'd first met Mike in 2002, walking up past the Nelson Cathedral to his hillside house to talk through the geology of the Red Hills. I knew only that the hills were part of New Zealand's Dun Mountain Ophiolite Belt, and that ophiolites were rare. They'd perhaps make an interesting three-day traverse within the longer New Zealand route and I'd wanted to know more. I'd offered up my notebook, and Mike had sketched in the origins of the belt. Then, as part of the route-finding for Te Araroa, I'd gone into the Richmond Range, three days walking south-west before I saw the Red Hills in the distance, glowing through a screen of trees, and as I got closer, saw the beech forest that recoiled at the foot of them. The Red Hills were not of this world. They looked like they'd flown in from Mars.

Now I was landing at Nelson to meet Mike again, and then I wasn't. The cloying white mass outside the window turned to cloudy runway grey, then the engines opened up again and the plane climbed to a holding pattern before setting course for Blenheim. By the time I arrived at Mike's house it was early afternoon and the tops around Nelson were still clouded in, but he was ready to go, and came out the door with a day pack and a larger New Zealand classic that he swung into the back of his Kia 4WD.

— Its motto is 'Mountain Mule carries the load'. That's fine, and so it does. But then, of course, you have to carry the mule.

The story of the ophiolites in New Zealand is an old one. Dun Mountain behind Nelson is 40 kilometres north-east of the Red Hills but part of the same belt. Its nearby slopes run blue and green with copper salts, and in the 1850s Nelson resident William Wrey had jawboned those colours into visions of a copper mountain. Wrey's familiarity with Cornwall's copper mines gave him some credibility, and adits driven into the hill did strike veins of copper ore. The Dun Mountain Copper Mining Company was registered in 1857 in London to mine the hill. Together with the province's coal and placer gold, Dun Mountain seemed poised to confirm Nelson's growing reputation as the mineral province of New Zealand.

By 1858, at the time of Ferdinand von Hochstetter's arrival in New Zealand and on through 1859 and his gradually expanding geological mission here, Nelson's Dun Mountain enterprise was stalled. The newly appointed mine manager of the Dun Mountain Copper Mining Company, 30-year-old Thomas Hacket, had arrived in October 1857, explored the adits and dismissed the 65-year-old Wrey's vision of a great copper lode as moonshine. He did, though, propose mining chromite from the hill. Out in the wider world, this newly discovered mineral was beginning to command a good price as a brilliant green replacement for arsenic dyes and for leather tanning.

Wrey had powerful supporters, Hacket had his determinations, neither had the evidence to stymie the other and the Nelson Provincial Government, hoping for a reliable scientific assessment of the mountain and for a wider appraisal of the province's minerals, put in a bid for Hochstetter's attention. Through to his final departure for Australia on 2 October 1859, Hochstetter would travel for two months through the Nelson Province and spend four days assessing Dun Mountain. He would later, in a diplomatic but decisive report, dismiss the copper as nothing more than small accumulated lenses, but he gave ongoing support to Hacket's assessment of the chromite. The Dun Mountain Copper Mining Company's shareholder money was finally put to use constructing a railway that carried the chromite down the mountain, to Nelson and shipment to international markets. Within a few years it was undercut in price by the rich ores of Baltimore, and other new mines, and undone by its own lack of reserves.

Dun Mountain proved a commercial dud, but a landmark for world geology. It allowed Hochstetter the honour of naming a new rock type, and in his *Geologie von Neu-Seeland* he described the rusty mountain, and the name he'd chosen —

> *Countless blocks of rock cover the slopes, and between them grow only low undergrowth and alpine plants, so that the dominant colour of the rock is only slightly masked by the vegetation. This rock is a highly remarkable occurrence of massive olivine, in great masses forming mountains, and is a truly eruptive rock so that it well deserves a name of its own, and the name dunite which reminds one both of the locality and the yellow brown colour of the weathered surface is most suitable.*

The rock was both 'massive' and 'eruptive' and for the next 100 years geologists would try to make sense of its origins. It had issued out of the mantle like any of the igneous rocks, but any such upwelling of white-hot rock inevitably spread heat through the surrounding strata, and changed that country rock in obvious glassy and mineralised ways — the so-called 'metamorphic aureole'. The point about Dun Mountain, and the Red Hills that lay beyond, was the lack of a metamorphic aureole. The ophiolites had come in cold and in the static world of the Old Geology that made no sense.

MARLBOROUGH SOUNDS AND THE RED HILLS

The ophiolites — from the Greek *ophis* (snake) and *lithos* (rock) — were puzzling rock suites. Ophiolites surfaced near island chains that had been forced up from the ocean, or within continental mountains that had been forced up from the coast. The American Harry Hess was one of a small group of ophiolite specialists who found that coincidence suggestive. He and other specialists argued over possible emplacement mechanisms, and over the melt points of the various ophiolite rocks, but coming into the 1960s a consensus had emerged at least on structure. Lifted up and presented in cross section, the suite of snake rocks might be five or six kilometres high and, starting from a base in the earth's upper mantle, would ideally have all or at least some of the following —

- Peridotite — a high-temperature, high-pressure mantle rock, rich in iron, magnesium, the olive-green mineral olivine and pyroxene. Variations of peridotite:
  — Dunite, which is dominantly olivine with layers of chromite and has yet to shed its pyroxene, and/or
  — Harzburgite, whose pyroxene has precipitated out in distinctive crystals, and/or
  — Serpentinite, the dominant rock within most ophiolites. The minerals in peridotite form at high temperature and high pressures, and as they're raised up they're out of equilibrium and turn into hydrous minerals, absorbing water and converting the olivine to serpentine, the dominant mineral within serpentinite. Serpentinite owes its distinctive name to similarities with the skin of a serpent.
- Plutonic rocks, such as gabbro, which form upright dike complexes.
- Pillow lavas — the most refined magma of all, fed from the dikes but billowing out on submarine surfaces and quickly chilled.

- And then sometimes a last layer, only tens of metres thick — marine sediments.

California had ophiolites, so did New Zealand. The Stanford University geophysicist George Thompson took note of that, and he took note of a further symmetry — the rock bands that lay either side of the ophiolites were also similar. Thompson arranged for the head of the New Zealand Geophysical Survey, Trevor Hatherton, to join Stanford staff in 1966–67 and shed some light on why.

The geology of the mid-1960s was changing fast, and the ophiolites would finally be a key part of the conceptual change between the Old Geology and the New. At the level of theory, geology advances on a set of rational fictions. Nineteenth-century geology had explained the mountains of Europe and America by a contracting earth that wrinkled like an ageing apple, but by the early twentieth century this was supplanted by the geosyncline, a model that acknowledged the vast amounts of sediment drifting off continental masses, and explained mountains by the principle of isostasy. It recognised the weight of offshore sediments, the heat and metamorphic conversion generated by deep burial, the plastic rebound from the mantle, some of it true and all of it clear and simple, scarcely different from the rise or fall of a ship loading or unloading its cargo at the wharf. The earth sank in geosynclines offshore and mountains rose, but the movement was mainly up and down, with folds.

Hatherton had the misfortune to land in California right on the years of transition from geosynclines to plate tectonics, but the new theory wasn't there yet, and he had only a year. He ventured forth, wielding the two primary subdivisions of the geosyncline model. His paper was titled 'Geophysical Anomalies over the Eu- and Miogeosynclinal Systems of California and New Zealand'.

Eugeosynclines were offshore troughs deep enough to produce hard indurated rock like greywacke. Miogeosynclines lay on the landward side of the eugeosynclines, still undersea but shallow, and produced a softer metamorphic rock.

Whether hard from the deeps, or soft from the shallows, the rock rebounded in million-year cycles up from the sea to form distinctly different bands of onshore rock, but the peculiarity that Hatherton sought to explain was the long strips of ophiolites that were wedged between those two rock bands. California had those long thin strips running for hundreds of kilometres parallel to the coast. So did New Zealand.

Hatherton flew inland from the coast, a hundred or so kilometres, recording the magnetic signatures of California, and then New Zealand's South Island. Beneath him, the separate rock bands yielded their distinctive magnetic signatures, the greywackes of the eugeosyncline, then the ophiolites, which were mineral-rich and made the magnetometer needle jump highest, then, further inland, the weaker signals of the miogeosyncline sandstones. The magnetic signatures he recorded across California and the lower reaches of the South Island were almost identical. It was confirmation of a particular type of Pacific margin geology and one that would interest other researchers as plate tectonics theory developed, but the theory wasn't there yet.

The prevailing theory suggested the ophiolites intruded at depth into the eugeosyncline, triggered by the syncline's descent towards the mantle, a hot undersea intrusion into what was already, at depth, a hot sedimentary pile. That might at least go some way to explaining why ophiolites, embedded as they were in continental crust, and bearing as they once had unspeakable heat from the mantle, nevertheless left no trace of the metamorphic aureoles volcanic eruptions routinely produced in their hot ascent through cold continental crust.

Hatherton challenged the idea of the ophiolites' intrusion

at depth, and his alternative suggestion glanced closer to the truth of it, but ultimately he was doomed by the wrong model. Hatherton suggested instead the ophiolite belt might pre-date the sedimentation. A possible modern analogy, he suggested, was the undersea ridge extending south of New Zealand to Macquarie Island. New Zealand's own aeromagnetic and bathymetric studies suggested a trough on one side, comparative shallows on the other, and the ridge between awaiting, as it were, some future role of dividing deep eugeosynclinal sediments from shallow miogeosynclinal ones.

The fictions serve until they don't, and geologists gather their evidence and squeeze it into the prevailing paradigm until they can't. Hatherton was enough of a good scientist not even to try.

'No reason for the formation of these ridges is given here,' he wrote, throwing up his hands in surrender.

Hatherton was back home as newly appointed director of the Geophysics Division of the DSIR when, in early December 1969, a watershed conference at Asilomar in California brought together the latest published evidence for coastal mountain building. The conference simply dumped geosynclines and their rebound as cause, and proposed a new model. Coastal mountains rose to accommodate the subducting plate at their feet. The appearance of ophiolites within those mountain chains was still unclear, but geologists at Asilomar rose to their feet to support a new concept: that the ophiolites were great cross sections of mid-ocean ridges escaped from the subduction process and impaled within the continental mass. They'd emerged white-hot at the mid-ocean rifts, but by the time of their intrusion into a land mass, they'd been travelling maybe 100 million years. They came in cold.

The cloud over Nelson stayed low and we abandoned the plan to land on a cirque below the Red Hill summit and tramp down. We'd be lucky simply to get to Top Wairoa Hut at 800 metres, and on that understanding the Robinson 44 flew in under the cloud, following the Wairoa River through steep valleys.

— Those pines down on the left, came Mike's voice through the headset. They're on the serpentinite and look at the difference.

Mike's finger was jabbing down on a commercial forest laid out in serried ranks below, tall and green except for a narrow band of forest trees that looked sick. The ophiolite belt where we were headed was whole tens of kilometres wide but here it narrowed to a corridor about a kilometre wide. Call it a hard row to hoe. No tree had ever evolved to cope well with ophiolites — the soil was too thin, and too rich in minerals. The magnesium, in particular, blocked the uptake of calcium, but when your seed fell on stony ground or some forester planted you across the boundary you had to make the best of it, and here they were. Within a lush pine forest, a population of indigent pines, stooped and yellow-headed, their trunks thin, their branches dipping with fatigue.

By now, the wider panorama of the Red Hills was starting to open up and the pilot, realising within the cross-chat on the intercom that he had a geologist aboard, asked the question that the Red Hills pose to any observer.

— The vegetation just stops up there, so what's going on? I always get the hunters asking me.

— It's a two hundred and eighty million-year-old bit of mantle caught up in the crust, said Mike, looking through the bubble to the big bald mass ahead. It's oceanic.

Below us stood Top Wairoa Hut, bright orange top to toe, except only for the black LandSAR number 513 painted on the roof. Its orange outhouse stood alongside. The Robinson set down above the hut, disgorged its passengers and their packs and peeled away. The hut was right on the Red Hills fault line. Native beech forest advanced to the far bank of the stream below and stopped there, a pretty selvage of ascending green layers afloat and happy on an afternoon breeze. On the hut side of the creek, the vegetation looked like it had eczema.

Two trampers were already ensconced in the hut, but we said only a quick hello, dropped the packs, and left for an immediate exploration, making our way down a steep slope to the Wairoa River and following the mashed rock along the fault.

— This is not soil, it's ground-up serpentinite, said Mike, rubbing it between his fingers and letting it drop. It's been reduced to pug by tectonic processes.

He grabbed a larger bit of serpentinite, a chunk of shining black and green rock.

— It's intact, but it's still tectonic movement that gives you these polished surfaces. A bit like beauty perhaps, it's only skin deep. If this was greenstone what you see on the outside here would go all the way through. True greenstone is more or less the same material, but it's down in what's called the Pounamu Ultramafics. It's been deeper, under the schist, while here the serpentinite has been under a lesser weight of sedimentary rock. Basically, it just hasn't gone as deep but, still, you can pick up boulders in the Wairoa River lower down which are semi-nephrite — not quite true greenstone, but they're good hard rock and they take a polish.

Serpentinite and greenstone. The difference was in the fabric, and Mike demonstrated by clasping his hands so they interleaved easily like a wigwam, but pulled as easily apart. That was serpentinite. Then he angled his hands to the fingers crossed at right angles. That was greenstone.

— Greenstone tends to have this coconut-matting-style fabric. The minerals are interlocking within it so when you hit it, nothing much happens. Which is why it's so precious to Maori. The greenstone is forever.

We went down to the river, and Mike picked up a dull brown

rock, held it in his hand, and hit it with his hammer. The interior fell open.

— Dunite.

My heart gave a small leap. You wouldn't expect that of stone, but the dunite had a history, and it was beautiful. Nor could you do better with a description of what was right then held out to me than to reference Hochstetter himself —

On freshly broken surfaces the dunite has a light yellowish brown to greyish green colour and shows a greasy to glassy lustre. Its texture is crystalline granular. The fracture faces are uneven, angular; granular and coarsely splintery; but on the individual grains it represents a fissility in one direction very clearly seen in small reflecting surfaces with glassy lustre.

Within that fresh face, Mike picked at numerous small black dots.

— Chromite.

He picked another rock out of the river. The hammer swung again, and a flake flew off. Mike picked it up.

— Argillite.

He ran the sharp edge of the rock along the hammer, and it scratched the steel.

— This is similar to what Maori would have used for adzes. Either off a big block like the quarries around the Rush Pool, or they moved up rivers and found boulders, hoiked them out and broke them down. Sometimes beside rivers like this you'll find the remains of a boulder, just a small pile of argillite chips.

The Wairoa's wide bed here was marked more by its settlements of boulders than by any river flow. As far as you could see back upriver the Wairoa was populous with boulders of every size, washed pell-mell out of the hills, or too big to move and simply excavated out of their matrix by centuries of river water, and Mike was happy amongst them. He moved off upriver and I watched him go, saw the gaiters and the plaid shirt and floppy hat disappear and reappear, noted the ease of his ascent, heard the *ching ching* of the hammer when he stopped, and pictured the alert enquiry as he bent over the exposures.

I climbed back up to the hut and talked to the two trampers. Their efficient movements around the hut marked them down as seasoned hikers, the man preparing a hot bath for the woman's feet and a compress for her sore knees. Little wonder. They'd been walking on rough ground from Hanmer and were now just two days shy of their house at Stoke, a journey of well over 200 kilometres. Mike came back in a while, dragging firewood to stack in the alcove for someone else's winter. We settled in, and I read the hut book.

The number of Te Araroa walkers who signed themselves through had roughly doubled most years, from nine in the 2009–10 year, to 24 in the 2010–11 year, to 40 in the 2011–12 year, 71 in the 2012–13 year, 75 in the 2013–14 year, and 44 from 31 October 2014 to 29 December 2014, the day we arrived, the tramping season right then only just beginning.

The Richmond Range and the Red Hills is a tough route, and the hut book listed its trials. Liz had slipped on an ice patch and 'hurtled onto rocks doing a 2 x 360°'. Coming up through beech forest north to south from mid-Wairoa Hut, others had been stung by wasps, or they'd been forced to ford the rain-swollen Wairoa seven times on the tramp up, and felt fear. They'd sidled past drop-offs, where the track camber tipped them outward above the drop, not inward. 'Dangerous!' wrote Paul. They'd reached the hut, opened their food, and watched the flies pour down the chimney with an amplified hum. Those who'd come through south to north from Hunters Hut were sometimes 'exhausted' and often they wanted more orange snow-poles to dispel the doubt that overtook you in misty weather. Dave and Clare wrote 'We ♥ GPS'. 'Sinking in snow up to mid-thigh,' wrote Boris after breaking trail in winter along the same high route, the beginning of a seven-hour slog that ended when he fell through the hut door here and would write 'du fond au coeur, MERCI!' Others summed up that same gratitude for the snow poles and the hut that hove in sight at snow poles' end, with the simplest, the strongest, the most eloquent word of all — 'Orange!'

Amidst this litany of cartwheeling, fording, slipping and slogging, I found the mea culpa I'd seen before in the huts. At the end of a gruelling section like this one someone would write, as someone had for March 2012 —

'I'm so sorry . . . for everything.'

It was signed off 'Geoff Chapple'.

In the morning we started early, ascending against the great red flanks of the ophiolite, and the track was fine until it disappeared entirely at the edge of the boulders. The Red Hills equivalent of South Island greywacke scree was boulder fields of red harzburgite with crystalline surfaces of pyroxene, sharp as tiny razors. Your boots couldn't slip on a surface like that, but nor, when they rocked underfoot, could you regain balance by steadying yourself against the promontories. We came through the boulder fields like slow dancers of the mosh pit, avoiding any touch of the razored world through which we moved, hands waving free.

Beyond the boulder fields I found a small rounded piece of quartz, then a more numerous scattering of them. Quartz doesn't belong within ophiolites. I showed the pebbles to Mike. Something had to have brought the quartz in, and he did a swift analysis of the possibilities. Any glaciers of past ice ages would have to come in from above where we stood, but that way lay only the ophiolites, and ophiolites didn't produce quartz. The streams hundreds of metres below where we stood could excavate quartz out of the sedimentary rock strata there and roll it along, but only downhill.

A bird could do the uphill. The quartz was rounded, without edges, and Mike concluded

— Yes. I think there's a high probability they're a moa's crop stones.

We stopped again amidst a wild field of dwarf flaxes, the same dry stalks of flax you might see elsewhere except they were thin, and their remnant seed pods, stripped by the wind down to a fibrous matrix, were streaming in the breeze, purely white.

— The minerals might be preserving them, said Mike. We did find during our mapping some very tough and resilient plants in the hills. The minerals get into the manuka for instance, and it takes a helluva lot to break them.

We looked down across tussock onto the Red Hills Fault. The beech forest below kept to its sharp boundary, flourishing on the sedimentary side, recoiling from the ophiolite side, but there was something else going on along that margin — a wide belt of bare land with gigantic boulders standing proud. They looked strange. They weren't brown or red, but black and grey, large and rounded, the shape of giant igloos. Some were sufficiently large to be crowned by scraggy trees.

— The melange, said Mike, pointing.

Although continually exhumed by erosion, the melange is long dead. It hasn't moved in millions of years, but it still preserves within itself the violence that rocked Gondwana 280 million years ago. To ask how and why is to step directly into the annals of that former world.

When you map the ophiolite belt north to Dun Mountain, and onward from there through D'Urville Island, it begins to look very long, but that's the least of it. Aerial magnetic surveys even in Hatherton's time tracked the ophiolite extending underground across Taranaki, Waikato and Auckland, up to Ahipara and out into the Tasman Sea. Though severed by the Alpine Fault, and offset 480 kilometres, the ophiolite belt picks up again at the Red Hills Range in Westland and runs on south-east through Otago and Southland and out into the Pacific.

But forget any placenames, whose present-day twinkling serves only to diminish the dim grandeur of it. The incoming oceanic plate of the former world is huge, anonymous, and as it overshoots the trench at the foot of Gondwana's proto-Pacific Coast and ploughs into the continental crust, it initiates a tectonic event so vast that the evidences of it will be easily discernible 280 million years into the future. The high face of the rogue plate butts against a mountainous coast thousands of kilometres long. To call it a collision is to seriously misreport the speed of it, yet no other word serves. Millimetrically slow but irrevocably sure, the impact of that collision severs pillow lavas and gabbros off the incoming plate on one side and rucks up sandstones and mudstones from the Gondwana coast on the other, and as part of that same sedate violence, the weak and hydrous serpentinite that shears up along the collision zone gathers these knocked-out bits and pieces and drags them down, block by block, 10 or maybe 20 kilometres.

The interface of that great Permian collision stretched for thousands of kilometres — the magnetometers of a distant future would attest to that — but of that length, only a few tens of kilometres would ever again surface. Two hundred and fifty-five million years on, in the early Miocene, the continental margin that holds within it the evidences of that relict collision will have separated out from Gondwana and sunk, but another trench entirely will have propagated across the sunken fragment and begun the uplift. An alpine welt will rise and bring with it the dunite and harzburgite remnants, and the jumbled assembly of the melange. Millions more years pass before Maori begin to wander amongst those exhumed monoliths, tap them to assess their uses, and report back to the tribe on the best of them, the highly mineralised mudstone they'll call pakohe.

We climbed to the saddle. Mike had gone on ahead, disappearing over the ridge into a stormy sky, and found shelter behind a big block of basalt. We broke out a snack. The sky opened briefly on a mountain and I got some further idea of the scale of the melange.

— That steep summit, said Mike, waving a cracker towards Mt Ellis 250 metres above us, is another chunk of basalt thrown up by the melange. The skirts below it are serpentinite.

The summit block, he guessed, would continue down 50 to 100 metres into that matrix.

We hunched behind our block of basalt, cloud blanked out the surrounding land, and I asked Mike to picture Gondwana. Geology calls it palinspastic mapping, unwinding the deformations within mobile space and sequential time to restore original landscapes, but Mike made the point: if change was continuous within the million-year cycles of geology, then which of those changing landscapes would you want to call original?

— Presumably, though, Gondwana was like most of the continents, fairly low-lying. You can get a fair idea of what the climate was at the time from fossil vegetation. You can work out where particular rocks were, like these basalts. They preserve the paleomagnetics like little frozen compasses so you can work out where the north and south poles were, and latitude, too, from the more subtle inclinations.

— You wouldn't be able to make a topo map, but you could pick the fold belts. Presumably also, Gondwana had big rivers, so you'd be able to place those. You could say there must have been a river draining out to deposit this particular sandstone formation. The thing is, of course, that the sandstone formation, which was deposited within a certain time bracket, will be limited, because much of it's been eroded off. So you'll only ever get fragments from one particular time.

— If you were able to check every fifty million years and had a look at the distribution of rocks you had preserved within that fifty-million-year time slot, you might be able to see deltaic sediments over here, and sand dunes over there, or rocks that clearly show they've been tilted so they must have been, if not an actual mountain range, then the remnants of one. Yes you could do that sort of thing, but if you tried to put all of Gondwana back together again on the basis of those fifty-million-year analyses you're still missing a large number of pieces, and if you're looking at just one fifty-million-year time slot you're going to have a helluva lot of pieces missing. That's the problem.

A tramper came out of the murk beneath Mt Ellis, a bright yellow storm cover on his pack, and he must have been easier to spot than we were as we moved out to intercept him, for he visibly jumped when we called out a greeting.

As it turned out, he'd had his 2014 Christmas in Christchurch and had taken the car up to St Arnaud where his wife was due to pick it up and drive on home. He, meantime, would walk home.

— And where's home?

— Stoke.

What was it about Stoke? I cited the two other walkers who'd parted ways from us that same morning, also headed for Stoke. The walker seemed unperturbed.

— Yes. I don't know them, but I see they're walking from Hanmer to Stoke. They're one day ahead of me.

He disappeared, but the wonder remained. I turned to Mike.

— Everyone we meet is walking to Stoke!

— It's not Mecca, said Mike. It's Stoke.

We waited on the weather, then started out again in the hope the cloud would lift, climbing the flank of the Red Hills ridge off to the south, hauling ourselves up on snow grass with Mike occasionally clouting some outcrop with his hammer.

— Good protoclastic harzburgite. And look, dunite, with thick veins of chromite.

On the strength of it, I suggested a joint stock company to mine the ore that started as a dyeing agent, then made prominent later progress plating the front of America's big V8s, and as an alloy to stop corrosion and harden the world's steel.

— Agreed, said Mike. We go in, float the shares, and we're gone again before anyone realises the economics are crap.

— That hammer of yours, I said as it descended again and clove another of the chromite veins that might make our fortune. Everyone else I've met has got an Estwing.

— This one, said Mike, weighing the brutal little club in his hands, is from a Dunedin foundry. For these rocks, you've got to have a proper hammer.

We climbed to the Red Hills ridge, Mike leading, until he stopped suddenly and drew his hiking pole across a strip of grey that lay flush with the contour of the ridge. The surface of it was broken up so it seemed nothing more than a slew of grey road metal bucketed across the slope, but like everything else in the Red Hills you had to let your mind settle, and maybe the company of a geologist helped, one who saw his world in four dimensions, who could wind this strip of grey back to an origin, and tilt it upright. It was a dike, the only rock in the world whose natural orientation is vertical.

— It's a gabbro or a dolerite, said Mike. A late intrusion. The ophiolite has already cooled down, so this is basically intruding solid rock. It's good you've seen this. Certainly this was active long before the ophiolite got emplaced into Gondwana. It'd be a late stage in the spreading ridge saga.

A late stage in the saga: this wasn't one of the multi-layered dike complexes at the centre of the mid-ocean spreading ridges, the ones that strike upward, that cool, that are riven from within by a new strike, that cool, that are riven from within . . . Repetitions over aeons that do not cease.

— We do have a fossil dike complex like that on the north side of Mount Ellis, said Mike. This one is a slightly younger intrusion as the sea floor, which here is harzburgite, has moved away from the spreading ridge.

A basement intruder forcing entrance through the floor with a crowbar. Geology texts often present sea-floor spreading as a smooth process, but it's smooth only within geological time, where mountains rise smoothly and as smoothly melt away. Maybe once every 50 or 100 years the dike complexes renew themselves at the spreading ridge or, like this one, sunder a flank, and keep the sea floor cranking along. Here's what I saw in the mist: a flat grey shadow maybe two metres wide that helps drive the world.

Red Hill had been doing a slow dance of the veils, offering glimpses in blowing mist of the big ice-lathed cirques at its foot, but as we walked on the white gloom closed down further, and I found myself studying the tight little populations of scree flowers underfoot, or the elaborate silver swirls of a prostrate *Dracophyllum* flat against the rock. Mike Johnston, geologist, was stooped over his own small vision. He'd spotted a patterning in the rock, and he shared it.

— You can see the polygon.

— I don't see the polygon.

— It's somewhat crude, but you can see how the different shapes have been brought into sharp contact.

I'd seen South Island ridges where broken schist is sometimes arranged into natural mosaics, I thought by ice or water, but Mike's polygon could only have been done by serious ice. The rearranged rocks were the size of broken concrete blocks and formed a series of touching polygons, each one about three metres across.

— The ice seems to juggle them and even stand them on end to get the best fit, said Mike. It's like beehives or columnar jointing. They can't form circles, so they tend to form polygons.

By agreement, we dumped Red Hill summit as a destination, but Mike was hoping for a glimpse into the headwaters of the Motueka River, and we went on up the broken ridge and reached the vantage point. With the precision of a scripted epic, the mists began to part and I gave an astonished cry. The soft light shone on the rough meadows of another mountain cirque at our feet, upholstered in red. It gleamed on the thin river that ran away below the cirque and vanished as a silver thread into distance. It glowed on the softly U-shaped valley that rose either side of the river. Mike looked into the depths of this lovely vision for a long moment, nodded to it, looked back at me and pronounced simply

— Glaciated.

We ate our lunch looking at the view and waiting for the curtain to drop. It did that, and more. Fat raindrops began to fall out of it, and we packed up and scrambled back the way we'd come. Down through the high standing rocks, crazed as they were with the freeze and thaw that afflicts hard rock at altitude. Personally I was happy with the fast boulder-hop of our descent alongside these standing pillars whose stiff angles and jointed columns suggested a watchful frieze of armoured knights. I went one side of a large outcrop, Mike went the other, I met him again on the other side, and he was stopped.

— I think we're on the wrong ridge.

The Red Hills ridge had its harsh cap of standing stones but it also had softer shoulders that beckoned you on to faster travel on broader slopes. Without any other landmarks, those misty shoulders were a siren call to just follow along. Mike was resisting the call. It was one of those moments where the territories you've been gliding alongside turn mute, the same mute that was camouflaged by your confidence, your jaunty inclusions of their rock stacks, their rock friezes. Silence. The mist. The alien rock stack. The alien rock frieze. Unease. Nothing much, but a possible growing point for something larger. I looked at Mike and he took out the topo map and an ancient GPS. We were heading down the wrong ridge.

We trekked back across the boulder fields. When you follow the wrong ridge, you diverge quickly, and we boulder-hopped fully 15 minutes before Mike looked down in satisfaction upon the mandala at our feet.

— Here's the polygons.

By now the mist was well down. The rocky outcrops ahead showed stark against the white, and now cairns stacked on the ledges stood out as small pyramids, or strangely balanced piles against the white backdrop. Cairns as clear as any set of lights guiding a pilot in, and I remarked on the clarity of the primitive signage leading us back to the saddle, back to the main track and the hut.

— The reason they stand out, said Mike, is that they look artificial. And they are artificial.

I asked Mike later what had alerted him to the wrong turning we'd taken. He gazed at me. The guy who, as a boy, had collected 400-million-year-old fossils in a rare Devonian outcrop up the Baton River, who'd won the Carty Cup two years running at Nelson College for tramping, who'd thrown up a proffered career as a Nelson dentist in preference for an outdoor life, and who'd been awarded the McKay Hammer for his mapping of the Red Hills and Dun Mountain. The guy who, as he mapped up the Motueka River Gorge, or down Beebys Creek and Porters Creek, wet as a shag at the end of the day from seeking the outcrops in those gullies, would hang everything out to dry, and for whom the great overlook of the Red Hills would act as a giant night store, sending its warm breeze down the gorges and gullies during the hours of darkness, drying the khaki shorts and shirt, the woollen socks, the Swanni and the Mountain Mule pack of the 1980s geologist. It seemed possible he loved the place.

— The wind on my cheek, he said. It was coming from the wrong direction.

WESTLAND
The alpine fault

Old New Zealand's geology was either the DSIR's down and dirty search for coal and minerals or, at its more exalted end, geomorphology. Coming up to the 1940s, Charles Cotton held the country in thrall to his *Geomorphology: An Introduction to the Study of Landforms*, then under revision for its third edition. Geomorphology was topography, and Cotton's pen and inks were interspersed too with his block drawings of stratigraphy underground, but there was little specialist attention to tectonics, which was stratigraphy on the move.

The Old Geology acknowledged an alpine uplift, but the prevailing explanation here too was the offshore geosynclines, that sagged under an increasing burden of sediment, put pressure on the mantle, and the mantle rebounded like a trampoline. There was no reason to look for a fault at the base of the Alps, and if there was no one looking, the fault was easy to miss, the strike of it softened by rapid erosion and concealed by dense forest.

New Zealand's Old Geology was strongest on its geomorphology, and Cotton's texts were popular and used by geology students worldwide. His intricate black and white sketches of the processes of erosion and land formation were nicely done, but some deeper insight into New Zealand's very violent tectonics awaited a more primitive geological consciousness than could be supplied by the prevailing academia. In 1941, the right man was on his way.

Photos of Harold Wellman in any group usually catch him glancing sideways, above and beyond any grinning human assembly. Smile to the camera — it simply wasn't in his nature.

He was smart, practical, argumentative and intuitive. He'd trained as a surveyor, but as the 1930s Depression took hold, he travelled south and went black-sand gold mining on the coast. Tiring of that, he began oddjobbing, including intermittent work for the DSIR. That contact morphed into a full-time job, and a suggestion that he fledge as a geologist by taking a degree. By 1940, in between fieldwork, he'd completed a Master's, but in August 1941, when the bus dropped him and Dick Willett, a fellow freshman geologist, at Hokitika, Wellman was still a raw recruit.

The nineteenth-century surveyor Charles Douglas had reported a mica lode near Paringa, and the two geologists were under instruction to assess its mining potential. They studied the Douglas maps at the Hokitika Lands Office and then, hitching south from the town and gazing left, Wellman saw continual slight offsets on the moraines, and notches in the hills. His training was minimal, but his eyes were fresh. He saw a continuing fault.

Suddenly curious, the two pushed into the West Coast's tumbled riverbeds and found the rounded cobbles of two dominant rocks — granite and greywacke. The rivers were excavating across a major unconformity. They picked up a lift on the back of a Model A truck, and as they rolled along, the passing landscape fell suddenly into an enlightening pattern. The solid low humps of granite lay on the western side of the fault, and the brittle ridges of greywacke climbed to the east. The fault was keeping pace with them as surely as the moon follows a child on a night-time street, and they mapped it at a steady 20 mph.

They assessed the mica at Paringa, then walked on to fulfil a larger self-appointed mission, past Haast, to the Jackson River, then followed the coast around to the Hollyford River mouth, and up to Lake McKerrow. They were then 100 kilometres due south of Haast, but still the fault scarp ran on, into the northern side of the lake and out the other side headed for Milford Sound.

It was the longest straight line on earth, and it bespoke pure tectonic power. Wellman named it the Alpine Fault and in his biographical notes described his approach to it, near Paringa: 'The broken schist looked like the result of explosions, or perhaps we were seeing the heart of old earthquakes.'

Christchurch 1949: twenty geologists pose for a group picture and the American seismologists Beno Gutenberg and Charles Richter stand at the front left, two obvious stars. As usual, Harold Wellman is all but divorced from the group, back row, off to the right. Everyone smiles to camera, but Wellman is not looking. He's gazing right, and westward, to an event horizon beyond the Alps.

The occasion is the Seismology Symposium section of that year's prestigious Pacific Science Congress. Richter and Gutenberg had already developed their magnitude scale, also a standard seismometer for measuring that magnitude, and backed by a Defence Department budget they'd salted those seismographs around the world. Richter's paper to the symposium listed global earthquake patterns, and he noted a 'vigorous deformation' on a 'circum-Pacific active zone'. That zone showed the San Andreas Fault across Southern California, and also marked in the Alpine Fault, but gave it no specific mention.

Wellman wasn't scheduled to speak, but he'd hinted big things and his impromptu lecture was well attended. As it began, he stood beside a large hand-drawn geological map measuring one metre by two, mounted on an easel. The map showed the South Island's distinctive rock bands in colour, including the reds of the ophiolite belt that included Dun Mountain and the Red Hills in the north, and which reappeared as the Red Hills Range in Westland. Part way through his lecture, Wellman suddenly slid Westland 480 kilometres along the Alpine Fault to match up the ophiolites as a single unbroken belt.

The room, from what little evidence does exist, was dumbstruck. Basic axioms of the trade were on the line. It was one thing for a man with scissors to sever an island in two — Wellman was mad enough. It was quite another to accept the earth was that mad. They were geologists, used to riding mentally up and down on the geosynclines, but what Wellman laid down in that room, that day, was almost incredible. Worldwide, no geologist had ever proposed such an astonishing offset.

Simon Nathan's book *Harold Wellman: A Man Who Moved New Zealand* recorded eyewitness accounts of a vivid moment. 'One professor clapped his hands over his head, taking it in at once and accepting it.' Not everyone did. Patrick Marshall, by then an *éminence grise* within geology, and that year to be awarded an honorary DSc by the University of New Zealand, rose to dispute it. Wellman tossed him the chalk and suggested he provide his own interpretation. Offended, the professor walked out.

The American reaction was not recorded, but in 1958, when Richter published *Elementary Seismology*, his standard textbook on earthquakes, he referred specifically to his experience in New Zealand and listed the Alpine Fault as a world-ranked master fault. Under the heading *Large Strike Slip?* he acknowledged Wellman's 'bold suggestion' of an accumulated 480-kilometre shift 'since the Mesozoic'. It was a cautious assessment. The Americans had evidence, most lately from the 1906 San Francisco earthquake, that the San Andreas Fault could slip sideways up to five metres at a time, but the conventional wisdom was that the fault's accumulated shift might total 40 kilometres.

But by 1958, Wellman was already far ahead of the scholarship, rampant with his own vision. He rejected the diffuse age range implicit in the Mesozoic's 250–65 million-year time span. He'd done more fieldwork along the fault, and he'd flown over the Haast–Milford section in a Cessna. In a 1959 paper he noted a '1000 foot horizontal shift of glaciated surface on the south bank of the Martyr River'. The paper ended with a simple logical sequence: if the age of the last glacial advance was 20,000 years ago, then the annual rate of movement to create the Martyr River offset was around half an inch a year, and at that speed 'the postulated 300-mile shift could have been accomplished since the Oligocene'.

In geological terms he'd lopped off over 100 million years, with huge implications for earthquake probability, but it was radical stuff and immediately put under challenge. In an authoritative 1963 paper, Pat Suggate, later to become director of the New Zealand Geological Survey and one of New Zealand's most highly respected geologists, acknowledged Wellman's work, then summed up all the research on the Alpine Fault and reiterated that the horizontal shift had been largely completed between 150 and 140 million years ago.

Wellman's response, in a 1964 paper, was to up the rate of offset to an average of one inch a year (25.4 mm) and a total time for the offset of 19 million years, an astonishing rate of travel, and accepted at that time by almost no one.

Coming up to the 1980s, with plate tectonics firmly in place, analysis of the fault would show the horizontal movement of the fault averaged 23 millimetres a year — very close to Wellman's estimate — and that it was taking up an estimated 61 per cent of the oblique pressures exerted by the Pacific and Australian plates. As to earthquake frequency, the fault was still mute, but there were ways of making it speak. Big earthquakes cause landslides, and the rivers issuing down onto the West Coast would cut through the detritus, leaving terraces either side of the downcutting that might contain dateable organic debris.

First off the blocks was a young geologist, John Adams. He prised fallen timber out of three river terraces and carbon-dated them by layer. His 1980 paper was the first to suggest a regular ticking of the tectonic clock. The fault ruptured on average, he claimed, once every 500 years, at around magnitude 8.

The range of the research was slim, but good enough to put the geological community and engineers on alert. Yet South Island seismometers recorded few shakes on the fault, so maybe it was aseismic, maybe it simply crept, the compression of the plates taken up by elastic strain across the Alps or absorbed in ductile zones.

In 1998, the Earthquake Commission commissioned Geotech consultant Mark Yetton to do a major study, and even in his initial desk work Yetton knew the aseismic school was wrong. Overseas studies suggested that the lack of earthquakes on a major fault did not signal quiescence. Rather, many of the world's big strike-slip faults were seismically evolved, their fault plane polished over time so that small and medium earthquakes dropped away and the big faults waited century by century for their single magisterial rupture.

Yetton extended Adams' river terrace investigation from three to 19, and his analysis went beyond the river terraces. He hand-dug two trenches across the fault, and machine-dug three others with a 15-tonne Mitsubishi MS140. He found unmistakable earthquake signatures, including liquefaction. Radiocarbon dating of both the river terrace and the trench material gave a broadband age for two large quakes, one within the range 1480–1645, the other 1700–1750.

Partway through the three-year study, Yetton met Andrew Wells, just then pioneering the new field of dendrology. Wells had a Scandinavian corer that extracted a four-millimetre 'straw' out of tree trunks. It was hand-operated, and tough work coring the trunks of hardy old matai or rimu, but the two persisted and the final results were astonishingly precise. In 1620 and 1717, the tree growth rings consistently diminished, even plunging in many cases from a regular average width of one or two millimetres to zero. Yetton backed up these dates with evidence for rupture length, using trench data and forest disturbance patterns. The 1620 event ruptured along 250 kilometres, he concluded. The 1717 event ruptured from Milford Sound to Springs Junction, more than 400 kilometres. The energy release implicit in those ruptures bespoke earthquakes around magnitude 8.

Studies that went on into the 2000s gradually pushed the earthquake record back to AD 1000, but in 2008, in deep bush, GNS scientists Kelvin Berryman and Ursula Cochran uncovered the geological equivalent of the Rosetta Stone. Not far from where Wellman had first confirmed the fault's passage down to Lake McKerrow, they began to expose an 18-metre bank, downcut by the Hokuri Creek. It was heavily vegetated, but if you cleared it with a spade the bank was a virtual barcode of earthquake sequence. Eighteen vertical metres of alternating dark swampy and grey sedimentary layers: the bank held one of the longest continuous quake records of any on-plate fault worldwide.

The two scientists carbon-dated the black organic layers, and their 2012 paper reported 24 ruptures going back 8000 years. The maths was simple — an average of one Alpine Fault earthquake every 330 years, the last one in 1717. They noted 'a fairly regularly repeating earthquake cycle' and insofar as straight maths can provide probabilities, they did their duty of predicting seismic hazard: a 30 per cent chance of a large earthquake on the Alpine Fault within the next 50 years.

If you wanted a headline from that research it was that the fault was ticking like a grandfather clock, and chiming on the hour, and New Zealand was within the last five minutes of the next chime. The big black hands were moving towards the next awful hour and everyone should beware, for being very old and with worn cogs and suspect springs, the old granddaddy sometimes skipped forward whole minutes and chimed early anyway.

What makes a geologist? In the late 1980s, Winstone's bluestone quarry at the foot of Maungarei was New Zealand's biggest. The quarry was sunk into the 10,000-year-old lava field like some extensive archaeological dig, sectioned out by high walls of black rock and terraces a kilometre and more long. There were trucks and front-end loaders moving to and from the crushers in one part of the quarry, and boutique industry picking away in another. A nine-year-old stands in one corner of the great quarry with her father, a stonemason who is seeking stone lintels for a barbecue. Somewhere distant from the main dust and the noise of the quarry the child finds olivine — a nest of the green crystalline mineral winking in the light. What causes the anomalous effulgences, the quirks and singularities of the earth?

Virginia Toy is regarded as one of the brightest young lights in New Zealand geology, winning a Geoscience Society scholarship on the way to a much-cited PhD on the ductile shear zones of the Alpine Fault. Her stamping ground then was Gaunt Creek, a remote cliff face accessible by track in from the highway. Virginia liked walking and fieldwork and if she didn't enjoy carrying big loads long distances, nonetheless that's what she'd done a lot of during the mid-2000s with a backpack of rocks, almost five kilometres out to the highway, sleeping in the back of her ute.

She'd gathered rocks, often in zero weather, often alone with a chamois bounding past, startled, and often some strange noise sounded upriver that you didn't know what it was. And what gave free rein to your imagination anyway was that you gathered something that in a very real sense was not of this world, but had been brought up from the depths, the ductile zone, where the rock was plastic and held within its shear bands and grain sizes clues to the forces that moved at the base of the Southern Alps. In the course of a four-year PhD she was learning to read the hieroglyphs.

Her rise within geology's predominantly male world was fast. When New Zealand needed to send a scientist to help analyse core samples from the great Tohoku earthquake of 2011, it was Virginia who got the job, and she joined the voyage of the Japanese drilling ship *Chikyu*, taking up station 200 kilometres out from the devastated Sendai coast.

Onshore the Buddhist and Shinto priests were still dealing with the spiritual distress that had followed the howling 40-metre waves, advising the sudden dead on their proper leave-taking, settling the eerie dreams of the living, exorcising the hauntings, and seeking to banish the long lines of drowned people that many saw trooping inland.

Aboard *Chikyu* it wasn't spiritual, it was practical. Its mission was to sample the undersea slip zone while it was still warm. The ship dropped guidance rigs seven kilometres to the ocean floor, then drilled a further kilometre of rock to reach the fault. The drill brought back its samples, a team of experts closed in. Virginia touched the five-metre band of lustrous clay that had lubricated the whole horrible disaster, and would help analyse it for a later paper. It was earthquake geology using the discipline's most advanced skills, and she was working with her international peers, yet the ship's fluorescent lights, the closed-circuit television, the piped air and the steel sleeping quarters kept her cooped up, and after the first weeks on-station she wanted out. She wanted to get back on the racing bike she rode 11 kilometres to her lecturing job at Otago University, back to the downhill speed, the wind in your hair. Back home to Spinel, her favourite amongst the chooks, and the sheep.

When I said I was going in by chopper to look at the Alpine Fault, someone asked me who with. Virginia Toy, I said, and the reply came, She'd be hard work.

Abrupt sometimes, rude by conventional measure, yes. The nose ring was a sign from her — as she put it — 'purple-haired and punk period' and she had a lot of attitude. She characterised herself as 'like the Dutch. They call a spade a spade. Blunt. I like that.' She'd done post-grad work in America, and it was less the science she found hard than the cultural demand that you had to be just fine, all the time. Their way of leaning forward with the anxious question 'Oh — are we alright now?'

Personally, I liked her. She was tall and thin, and she knew she was thin, fretting sometimes that she simply burned food too fast. She'd once put me up in the sleepout behind her Waitati cottage, just north of Dunedin. She'd been chopping wood when I arrived, cooked me a meal of local fish, and we'd finished up reciting poetry at each other. She quoted James K Baxter's early poetry 'Nor did I ask' with its fierce quatrains of a lost first love and the steely strength, unasked for, that followed —

Nor did I ask to walk
stronger, where no rocks bleed

I dredged up a remembered poem from the Australian Les Murray, one of the few great poets to tackle geology. The poem ended with a summary of tectonic stress building to the instant of disaster —

If coral edging under icy covers
or, too evolutionary slow

for human histories to observe it, a low
coastline faulting up to be a tree-line
blur landscape in rare jolts of travel
that squash collapsing masonry with blood

then frantic thousands pay for all of us.

It was a feat of memory, but I stumbled at the ending, took two cracks at it, and started in to say I'd spoiled the deadly kiss-off line.

— Not at all. You did it well, and I enjoyed the sound of that granular slightly croaky voice.

Virginia had no filters. She saw the world with extraordinary clarity, called it as she saw it, and said of her own profession

— Most academics have mild forms of autism and that means they are overly sensitive to some things environmentally. They sometimes find it hard to interact with people socially and they find it hard to read people's expressions as other people do. But it also means if someone drops 500 matches on the ground, or 501, they can switch off everything else and focus on that sound only and therefore they can tell how many matches fell. Or they can sit down and become so involved in their work that they get something done, like I do.

— They might be difficult, but what do you do? Homogenise it so everyone is fine and okay and don't have ups and downs and little aspects that are not the same as society, and then how boring do we get? I'm all for crazy people.

When I retired to the sleepout later, the adjoining pit toilet installed in a red telephone booth had a six-panel cartoon pinned to the wall headed 'Why stick people became extinct'. There were two stick people. In the first three panels, they meet, kiss, then conjugate. In the next three panels the flames begin at the groin, spread, and presently there's just a pile of ash.

Virginia was smart, engaged with the best earthquake science in the world, loved wild places, rejected social convention. If you wanted to see the Alpine Fault, she seemed like the right person to take you there, the closest thing in the country to an Alpine Fault avatar, and as it happened she was going in for a three-day July field trip and invited me along.

In the exchange of emails as four of us arranged the gear we'd take in, Virginia listed the necessary equipment and permissions: field books, compass, GPS, hammer, chisels, marker pens, sample bags, paper tape, tinfoil, a sampling permit, the latest topo maps.

Without geology equipment myself, I suggested I could bring a small and light travelling guitar, and the reply came back.

— I can survive without the guitar.

We packed into a chopper at Haast and headed towards the Alps. The land dropped away below, and you could see great areas of pakihi, and the tall skinny forests of the flood plains holding wisps of mist, the big granite bumps that had intrigued Wellman long ago, then the beech forest on the uplands which rose steeply towards an alpine snowline.

The small team comprised Virginia and two Americans, Steve Kidder from New York's City College and Josie Nevitt from Stanford University. Overseas geologists come to the Alpine Fault because it's one of the few active plate boundaries that brings its own roots to the surface.

I asked Steve what it was he'd most want to find on the present expedition and his answer was pink chert. The chert was a hard sea-floor rock, laid down in the ancient seas by the mineral skeletons of minute protozoa, but if you were lucky, you might find a pink translucent band of it ascending from its sea-floor origin towards the summits of the Southern Alps.

We dropped through a hole in the clouds, unloaded, tied a wide blue fly to a scraggy line of trees as camp central, then pitched our tents separately. There was no geology that day. The rain fell, the WhisperLites hissed under the fly as we cooked our meals, and we went off to the tents soon after dark. I'd chosen a site down the valley away from the others, near the junction of Robinson Creek's two headwater streams. I picked my way down by torchlight across the tumbled rock and lichen on the valley floor, opened the tent zip to find a small lake, bailed it out, and settled down for a cold night inside two sleeping bags, upslope of where water was dripping through the tent seams.

Rain tapped on the tent, Robinson Creek roared like a train, and I could hear the occasional donging of a rolled rock. I went to sleep listening to the ongoing rush of the creek, comforted in its persistence, then awoke an hour later, made somehow anxious by the same ceaseless rush, and a larger stillness. The Robinson train was only a small train, but it didn't stop. Little by little it was transporting bits of the valley, pulling the place down, and the valley hung over it, fractured and silent. I thought of earthquakes, and in the morning, Virginia said, apropos of nothing at all

— Yeah. I thought there could have been an earthquake last night.

She wasn't joking. If Virginia thought that, then it seemed possible the fault was thinking the same thing. Everyone knew it was only a matter of time.

— The valley isn't that pretty, said Josie, looking around, and she was exactly right.

The high valley rim that stood all around us was edged by a Gothic lace of dead or half-dead trees where kaka sat and watched, and called down a harsh call. The walls were eroding away, hanging whole trees upside down like natural distress signs. We climbed steep and sliding piles of debris to reach pillars of actual bedrock, but the bedrock itself wasn't normal. At first sight I had a pleasing flash that this might be the same shattered bedrock where Wellman had his own salutary encounter, for Robinson Creek lay on the old Paringa to Haast cattle trail, and in 1941 that was the route Wellman and Willett had taken to walk through to Haast. 'The broken schist looked like the result of explosions, or perhaps we were seeing the heart of old earthquakes' Wellman had written, and here the words were full of meaning. The pillars and cliff faces around us were the outcrops I'd already seen many times in the South Island with their layers of schist and white quartz, and occasional thin dark layers of argillite, but here, though the stratigraphy was intact, every layer was broken into component pieces no bigger than grey Lego.

— This is just what rocks look like when they ride along on a fault for a while, said Steve. I've seen the same thing on the Simeon Banks in California. Blocks that have been transported several hundred kilometres on a fault. They're shattered, but they're all still in place.

— There must be a name for this?

— That's just the fault damage zone, said Virginia. Okay? So yeah, the fault slips during earthquakes and the shock waves cause in situ fragmentation.

The Pacific Plate, the largest of the world's plates, is three times the surface of the visible moon. The plate grows from the linear upwelling of molten rock along the South Pacific Rise, and jostled by other expanding plates, constrained by the logic of spherical geometry and a finite planetary surface, there's nowhere for the distant rim of the plate to go but down. So to the famous Pacific Ring of Fire — the plate descending at its edges, and volcanoes stippling the rim — but the South Island deals with the incoming pressure differently. Its crust is too light and too thick to be subducted, and if you can't go down, you go up. More than that, the greywacke crust on the Pacific side is contending against the granite crust on the Australian side, and if it rose steadily to respond to the pressure, still it accommodated the expansion of the Pacific Plate also by the third and last means of accommodating growth on a finite globe. By slippage. And though the Alpine ascent was the most visible snowy triumph of the island's tectonic encounter, still for every metre its Alps were riding up, they were sliding sideways by 20 metres, a protracted glancing collision.

The crust rose and slipped and either way it broke along the Alpine Fault in the roar of its recurring magnitude 8 earthquakes, but Steve wasn't dealing with those brittle crustal events. He was collecting thin sections of the rock geologists called mylonites, rocks that had been under similar strain to the strains that produced earthquakes, but which had accommodated the strain by non-brittle mechanisms. The Alpine slopes that rose above the fault were chock full of mylonites. Every whorl of white quartz in an Alpine rock bespoke a mylonite. The Southern Alps had risen something like 20 kilometres since their uplift began 12 million years ago, but it was long enough that the mylonites from those depths had been carried up to become part of the brittle crust. They brought with them a plastic memory that was open to the kind of analysis Steve wanted, but you had to pick the right ones.

— My general classification, said Virginia as we stood beside the rock, once you get to the point where you've got a craziness in the foliation, or where you can see quartz bands like this, you're getting into mylonites. If you can even remotely see the shear bands, then that should definitely be called mylonite.

— There seems a lot of difference, said Steve. There's several older horizons.

Grain size in the quartz, that was Steve's take. He'd gathered over 200 thin sections from top to bottom of the Alpine Fault and was shaping a thesis that stress levels were high and the quartz grains small where the uplift was slow, and conversely the stress level was low and the quartz grains large where the uplift was faster. In sum, the greywacke was rising more slowly at both ends of the fault than around the middle. The foliated greywacke there was — in Steve's words — rocketing to the surface, bringing its heat to shallow depths, delivering the hot pools everyone had always found odd, for there was no volcanic heat anywhere close.

Grain size sounded simple enough, but if your sample was going to yield information on the Alpine Fault, you had to be sure you gleaned it from the right tectonic event. To tie grain sizes to the Alpine Fault, you had to identify the right lineations — the geologically recent pull of the rock at ductile levels that was an Alpine Fault pull, and not some Zealandia pull, or a Gondwana one.

It was something close to a dark art to get the right lineation, and you needed the right one or you'd cut your sample the wrong way. Virginia was the go-to woman on lineation and mylonites, and she worked to set Steve right on the samples, standing right up against the outcrops, in the rain, compass in hand, a black brolly obscuring her head and shoulders, entirely focused on a grey wall of rock.

— Bags! said Virginia, deciding on a good sample from the fractured wall, juggling compass, brolly and rock pick, her hands full, and demanding instant action from the geological assistant's spare pair.

— There might be some in my backpack, she called over her shoulder.

The tall woman in a blue Macpac jacket, black brolly upraised, face to the rock, a figure quite as strange as the Magritte image of the man with the bowler hat, and an apple in his face.

Sometime later, with a happy cry, within an outcrop at the top of the valley, overshadowed by the great walls of broken mylonite behind, Steve found his band of pink chert. We left him to it, and walked through the wreckage of scattered trunks and strewn boulders, back down the valley. Virginia had some of her own work she wanted to do. She wanted to sample the fault itself, and, strangely, to that moment no one had mentioned the actual hub of it all, the thing that had kept me focused for weeks, and some of the group for years.

Down past the tents. I knew where I was going, but not exactly what I was going to see. I'd already twigged the change in landscape at the bottom of the valley. I'd lifted the tent flap in the night, looked out at a crescent moon, and seen the contrasting silhouettes of it. I'd stuck my head out of the tent in the morning, and seen more of it. As the two headwater streams of Robinson Creek joined forces there and gushed away downhill, you could see that it had cut down differently on two separate substrates. Beside my tent it rushed along the bottom of a relaxed V, but then it hit mashed gravels and the cut turned to a U, a narrow gorge with steep cliffs rising either side. The face of the cliffs was knobbed with boulders like a climbing wall 30 metres high, but you'd have been foolish to attempt any ascent there. The boulders were covered in moss, and instinct told you that every handhold on that soft wall would come off in your hand. It was the point at which Robinson Creek left the Pacific Plate's hanging wall and crossed to the Australian Plate's footwall.

Past the tents, on towards the bottom of the valley, and across the creek. I clambered onto a boulder, sat down and looked at it. At the very start of those high gorge walls, top to the bottom, the Alpine Fault slanted down. Big and blue, a smooth band almost two metres wide set across the green cliff at something close to a 45° angle.

We were there to work. Virginia wanted samples from the black rock that bounded the fault, and she assigned the procedural tasks she needed to fix them in their exact alignments for analysis back in the labs. Josie would remain on the far bank and, from a fixed position, record the heights of each sample with a laser rangefinder. She'd stay this side of the creek and write the strike and dip numbers Virginia called back over her shoulder. My job was to stay close to Virginia on the slippery wall, to take the raw samples she handed back, to wrap them in tinfoil, tape them, and write the numbers on the tape.

Virginia boulder-hopped across the creek to come up on it, and I jumped after her and landed at the foot of the cliff. I looked to my boots, seeking to brace myself against falling back into the water, and as I stared down saw that the bustling creek was pulling at the fault itself, stealing it in every moment, and the fault was helpless against it because it had no structure and was dissolving in a continuous swirl of blue-grey sediment and curling away downstream.

Virginia climbed high, not on the fault itself, which was as slick as a playground slide, but alongside it, on a darker rock that was almost as hard to ascend because it gave way constantly underfoot. Josie put the red spot on the wall, Virginia worked with a compass and inclinometer and called back the numbers, then cut out the samples one after another and I reached up from below each time and took the proffered sample. You would hardly have called it rock. It was a slice of black blancmange, and the only way to preserve its squidgy shape was to wrap it tight in tinfoil and strap it around with paper tape. As I climbed up and climbed back, as I trudged back and forth down by the creek, bunching each new sample with its mates, the black rock underfoot turned into a sludge that overtopped my boots.

When Virginia did her dip and strike, she was concentrating hard. You didn't talk to her, but later, back in camp, she spoke.

— You saw the difference, right? Steve's stuff is ductile shear zone. He's one of the few people doing really good still extremely topical and useful quantitative research and structural geology on ductile type rocks.

— What I was doing up on the wall was the brittle stuff. It's ugly rock to work with and that's partly why people like the ductile work. You can go and knock out this beautiful rock. It's got this lovely boudinage or lozenge or whatever in it. You can take the sample home easily. You can cut it up. You can look at the beautiful mylonitic fabrics in there.

— They're lovely to deal with but it's the ugly stuff that offers us the insights into earthquake mechanics. And you know earthquake mechanics is only important because it impacts society, like Tohoku, right? If you are more interested in why we get earthquake ruptures and what characterises those, and what we can learn to help us mitigate the hazards to society, then we need to be looking at the brittle zone.

On the third day we folded the tents and there was ice under my ground sheet. We choppered back out, got into Virginia's hired truck, and drove along the roads of southern Westland. The thin trees of the floodland forest stood straight up alongside the road, and their high green canopies tightened over the top of it. We were looking down the long tunnels Virginia had seen as a child, and she said

— I ended up working on the West Coast mostly because I loved the West Coast. I was taken here when I was aged thirteen by my parents. A Guide camp jamboree which I didn't enjoy much, but I did enjoy driving down these avenues of trees in the rain. Something about this environment makes it one of the most beautiful places on earth. There may be other places, but as well New Zealand has more trees, more rain, more light. I always feel most comfortable with the feeling of being in your own land, where you were brought up.

We drove across the Waiho River Bridge, into Franz Josef township.

— The Alpine Fault runs right across the highway, said Virginia. This bump in the road, just there right? The fault runs right through the middle of town.

Old New Zealand was either incredibly gung ho, or simply innocent of earthquake hazard. The nation had placed its prime minister's residence within metres of the crush zone on the Wellington Fault, and Franz Josef, the town that arose like Topsy as the tourist gateway to a famous glacier, would be the first to go topsy-turvy in an Alpine Fault earthquake.

I'd talked to civil defence planners back in Christchurch who'd calculated that a magnitude 8 earthquake on the fault would rupture the stopbanks on the Waiho River, flooding the town. Nor was that all. The fault famously bisected the forecourt of Allied Petroleum's gas station, and there was a 24,000-litre storage tank sitting in the fault's crush zone. The CD planners had two scenarios for the tank. It wasn't going to explode, but it was likely to rupture. If the tank ruptured after a dry period and ground water levels were low, the petrol would stay in the ground. If the ground was wet, the petrol would float to the surface and the fumes would drift across the town.

We were dressed against the cold in beanies and Macpac jackets, and we headed out to the glacier. Virginia walked at speed through bush, and we emerged into a wide flat valley with gigantic standing stones. I remembered something Steve had said, back at Robinson Creek — that stresses were lowest where the uplift was fastest. Here, near the middle of the Alpine Fault, the stress was lowest, and for whatever reason, the feeling of thoroughgoing destruction was absent. Maybe, surfacing twice as fast in this region, the Franz Josef rocks had that much less of the Alpine Fault's recurrent shaking. Maybe the glacier, once a much heavier and longer tongue of ice, had simply cleaned out the damage zone. Whatever the cause, the glacier's retreat had left behind a spacious, almost civilised landscape, polished at its edges into large smooth pillars of stone.

I turned off the main track to have a look. The stones stood strong and tall, striped by white foliation lines that emphasised their height. Those white quartz stripes had been squeezed out of the greywacke in horizontal depths aeons ago, and now, uptilted to 90° against the Alpine Fault, they pointed vertically to the sky. The thin stripes of the grey schist and the thin white layers of foliated quartz were as closely and regularly interlayered as pinstripe.

Ice always leaves a totalitarian stamp on any landscape. I'd seen the forceful architecture of the Emperor Ice many times, planing the softer South Island valley sides into the distinctive slope of a simian forehead, or the harder hanging valleys into steep walls and vertiginous drop-offs. Yet here, I thought momentarily, the emperor had turned his valley over to a more democratic regime of aldermen, who stood there, hands behind their backs, watching the tourists troop past on their way to the Franz Josef Glacier, conservative ductile gents wearing their well pressed pinstripe suits.

Virginia looked straight past the pinstripe, less interested in the old foliations than in the newer and thicker quartz veins that sometimes flashed across at an angle. Those held the clues to the more recent movement — the fat white boudinage, or the thin white pinch and swell veins that opened the way down to analysis of the ductile zone, and the creep of the plates. They were beautiful evidences, in an astonishing part of New Zealand, and Virginia turned from those to the tumble of white blocks in the distance, and waved a dismissive hand.

— Yep, there's some ice, right? This massive frozen bit of water that everyone comes to look at. Well okay. It's a fairly nice analogue for the deformation of rock, but honestly. What's cool about this location? Gee, a little bit of ice, watery stuff, retreating, hardly even visible up the valley, or the Alpine Fault? We're standing more or less on the boundary between the Pacific and Australian plates and the tourists are like — the what?

We went on up anyway, and watched a big chunk of ice fall off a melting arch, but she was keen to go. Before we left Franz Josef, I went into the service station to chat to the woman behind the counter.

— Well frankly, she said, it's either going to go or it's not going to go. I'm not about to pack my bags and leave town. If you're going to have an event like that then there'll be more to worry about than this little petrol station.

We drove on north, stopped on the highway just short of Whataroa, and walked the six kilometres in to Gaunt Creek. This was the place Virginia had done her original research. The fault stood out now as a greenish band on the cliff above Gaunt Creek. I photographed my hand holding the two plates together. Good for bragging rights on social media maybe, but after the damp lustre of the fault at Robinson Creek, this one was a husk of dried-out clays and fluvial gravels. What I'd most want, I said, was the actual interface of Australian Plate granite against Pacific Plate greywacke. Virginia couldn't quite fulfil that order, but after a brief search, she handed me a rock with a brown band through it.

— Here's a beautiful piece of friction melt rock from deeper in the crust. You can see a large band, probably some crystals in it. An earthquake occurred. It must have slipped rapidly.

Earthquakes start in the seismogenic zone at depths of between 5 and 15 kilometres. Deeper than that and the minerals are already starting to flow, shallower than that and the fault is clothed in clay minerals that can't store the stress. The clay-rich bands would rupture when the time came, in a follow-the-leader way, but the rock I held was a small 10-centimetre wonder that recorded hard-rock fracture from the seismogenic zone deep underground.

The rocks had once slipped so fast against each other that their minerals melted to a brown band. Volcanologists call their glass-rich basalts tachylites. The brown band within the rock I held looked similar, but was not volcanic. Hence geology's name for them, pseudotachylites. I called my one simply a fossil earthquake. I said

— These must be New Zealand's most exciting rocks?

— They helped get me involved, said Virginia, and now I wish everyone would stop asking me to show them pseudotachylites. They don't contain a lot of information about large seismic events. What they are, they're great things to get people interested in what I do, but mostly I don't give a flying fuck about them. They're one of those communication things.

The shipping container a few hundred metres away housed instruments that monitored the fault at a deeper level. She opened the doors and we peered in, but it was just fibre-optic cables leading into a capped hole. Virginia was a co-director of this 2011 drilling programme to sample the fault core at 150 metres depth. Early in the programme, a torrent of water had swept fractured rock from near the top of the drill hole into the hollow bore below, and presented geologists with their most dramatic finding. The fault was an impermeable seal that separated high-pressure fluids within the Alpine Fault's eastern hanging wall from lower-pressure fluids in its western footwall. The Alps rose above the drill site as a vast header tank.

— We've put seismometers down there, and pressure and temperature sensors, said Virginia, but fluid pressures in the hanging wall of the fault are high. The water sits on top of the step, so it's hard to get accurate measurements. To escape those effects we'll be going deep this year at Whataroa, about eight kilometres north of here.

This is how it begins. She's co-ordinating the entire science team at Whataroa, so she's drafting schedules until 10 pm. At 2.45 am she's pedalling in wind and rain the five kilometres to the drill site. As she crosses the Whataroa River Bridge, she can pick out, within the floodlit arena away to her left, the drill-rig's tall derrick, the trucks, the racks of drill-hole casings, and the makeshift labs she's so proud of. For her, this bright cynosure in a dark valley makes every effort worth it, for the science is thrilling.

They'll drill down 1.3 kilometres, intersect the fault, and pass right through it to the Australian Plate. Webster Drilling and Ecodrilling are both experienced New Zealand companies who combine their skills for the task, but they call it a greenfields project. Nothing like it has been done in New Zealand before. The project pushes drilling expertise — and geology — into the unknown.

The 120 scientists who're involved in the project haven't seen anything quite like it either. A geophysics team will regularly lower slim instruments into the deepening hole to log heat, pressures and strata. Another team will analyse the rock cuttings brought back to the surface from the fist-shaped PCD drill bit of open-hole drilling. Finally they'll trade the brutal PCD bit for a diamond-impregnated coring bit that preserves within it successive cylinders of the rock it passes through. That schist, clay and granite noodle will be hoisted to the surface and piece by piece will provide a pristine record of the fault structure at depth, the Pacific Plate one side, the Australian Plate on the other. And in the final stage the geophysicists will insert into the long thin chamber left by the coring bit a monitoring string to measure day by day the seismic tremors on the fault, the fluid pressure, temperature and deformation. The teams are setting out to find how the fault changes year by year, and if they monitor finally up to and into the black heart of a magnitude 8 earthquake, if they find distinct markers prefacing the event, then that's at least a glimmer, and maybe even a grasp, upon the grail of all earthquake science — prediction.

Here's how it proceeds, over months. The drill hole gets hotter quicker than anyone imagined. An artesian flow at 236 metres that's warm enough for your spa pool, and at 400 metres a rock temperature of 56°C. By the time the hole reaches its final depth of 1300 metres, the temperature is expected to be 190°C. Short of the Taupo Volcanic Zone, nowhere else in New Zealand has that kind of heat. In the Whataroa pub, the locals talk up a geothermal prospect, and the scientists try to explain, even one to another, how it could get this hot. Alpine Fault rock is rising fast enough to retain its heat to shallow depths, that's already known, but by itself cannot explain the Whataroa phenomenon. Maybe, under pressure from that huge hydraulic head they encountered first at Gaunt Creek, the creep of subterranean Alpine fluids along permeable pathways carries and concentrates the heat into superhot hubs under the Alps. Which hubs — so they speculate, beer to hand — may sit naturally in the valleys, in the Whataroa Valley, but other West Coast valleys too.

This is how it ends. At a depth of 890 metres, to provide a strong and steady base for the diamond corer as it spins on towards the fault itself, they must line the long and curving stone pipe they've created with cement. The hole is lined already with steel casing, but there's a small gap between the raw diameter of the drill hole and the smaller diameter of the steel casings — the annulus. They'll pump concrete down the middle of the casing to the base of the hole, and from there it'll backfill up through the annulus. They call in the concrete trucks, and begin to pump the concrete slurry.

I'd followed the deep drilling project closely by way of a regular blog put up by the co-leader of the project, Rupert Sutherland. A great read, with drama all the way, and team leaders talking to cam on their work, but on 15 December the blog headline read 'Calamity'. Unknown to the team, even as the concrete slurry was pumped down the hole, the casing string at 436 metres depth was already broken and offset.

In January 2015 I rang Virginia. She'd come back up to Auckland to be with her folks over the Christmas holidays. She rode her dad's racing bike down the East Coast Bays, and leaned it outside a Devonport café. She bought a cold drink, I ordered a coffee. We went to a pavement table, and Virginia sat there with her usual acuity critiquing the locals. She was having fun. The 40 tonnes of misguided cement that had buggered the deep drilling project wasn't weighing as heavily on her mood as I thought it might.

— Drilling projects are just about sleeplessness basically, she said once she'd settled in. They're about screwing up people's daily schedules. I'm not kidding, eh. The young post-grads who have the overnight shifts, they're like zombies walking around.

She was upbeat about the chances of another try. They'd get there. As drilling projects go, this one was cheap. And the high temperatures they'd encountered meant almost certainly the seismogenic zone was relatively shallow, and within reach of a future probe.

One further snippet from Virginia that morning opened the gulf back to Steno, and Hutton, and Cuvier, and Lyell. The drilling rig at Whataroa had been under transfer from Ecodrilling to the fundamentalist Christian Community at Gloriavale near Greymouth. The male Shepherds who ran Gloriavale had prayed to God, then in 2014 had approved the purchase of a rig capable of drilling for West Coast oil. A couple of Gloriavale's drillers were on hand at Whataroa to get familiar with their new rig.

And so, in that dark valley, to the history of geology redux. The Gloriavale drillers, both highly skilled in their work, defended a world that was 6000 years old. The facts of a Garden of Eden at 4000 BC, and Noah's Ark at 2000 BC, were taught in Gloriavale's school, where you might also learn that Noah's problem with housing the dinosaurs could be logically solved. He took aboard only the babies. Gloriavale's drillers on one side, and sleep-deprived geologists on the other, the crazy zombies who believed the world was 4.5 billion years old.

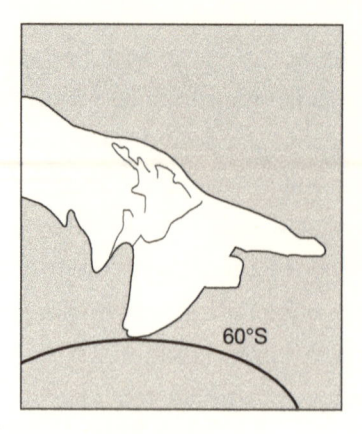

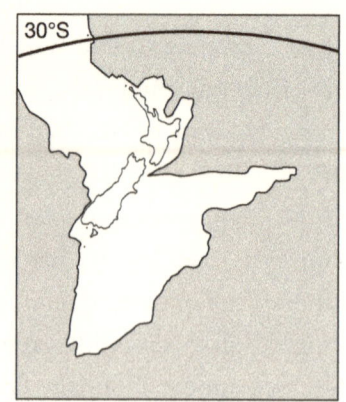

The Alpine Fault hadn't just cut the South Island in two. The same geophysical surveys by satellite and bathymetric data and dredging that established the boundaries of Zealandia also showed how the propagation and subsequent mix of subduction and transform faulting along the Pacific–Australian plate boundary was moving the entire sunken continent steadily north.

Zealandia, that is, as the Princess of Squirm, its elastic boundaries changing shape under the implacable forces of a jostling spherical geometry, but over the past 20 million years its waist had begun to narrow along the Alpine Fault. If New Zealanders talk routinely of the separate North and South Islands, Cook Strait in between, then it's just as routine now for local geophysicists to reference North and South Zealandia, even if, in the present era, they're separated by a band of clay just a few metres wide.

The intense interest Richter had shown in 1949 to Wellman's bold proposal was fired in part by knowledge of California's own San Andreas Fault. As would become clearer in the years after 1949, the seaboard master faults were rare worldwide because, in the course of time, the transform movement — that succession of magnitude 8 quakes — would separate the land mass entirely.

In the present era that schism is not complete, but it's under way. Stand inland of the San Andreas Fault, face towards the

Pacific Plate, and you're looking across a dextral master fault that's moving San Francisco and a large part of California steadily north. Stand in Westland, face towards the Pacific Plate, and you're looking across a dextral master fault that's moving the Southern Alps, Christchurch and most of the South Island steadily south. If you take the horrors visited upon San Francisco within historical time and Christchurch within just the past few years, and plot them on the wider gyre of geological time, then the tiny pointillist agony of those cities is underlain by this: New Zealand and California are bound to the same lithic wheel, 11,000 kilometres apart and moving steadily anticlockwise at around 25 kilometres every million years.

SOUTHLAND
Terranes

The New Zealand Stratigraphic Lexicon has on its books something over 7000 names for different rock strata and formations.

Find a formation, write a paper that's accepted by a reputable geology journal, peer reviewed, then published, and the chosen name for your formation gets logged into what's commonly called StratLex.

The lexicon has been built over many decades from the bottom up. Geologists have continued to fire off papers that bequeath a unique name to a rock formation of local interest, but as geology's understanding widens, the local outcrop may turn out to be only one part of a more extensive formation. The local names may then become simply clutter, and their persistence within StratLex often a barrier to efficient research and understanding.

Cast off or unused names accumulate within StratLex. Some fall by the wayside through disuse. No one now mentions the Pahi Argillaceous Limestone, the Pairatahi Limestone or the Pokapu Limestone. They were all once, and some still are, discoverable within StratLex, but all are now subsumed by the widely acknowledged Mahurangi Limestone, which is regional in extent. Such passive replacement is painless, but from time to time also a geologist with sufficient authority — a cleaner —will enter the labyrinth and actively recommend the removal of names.

At the cleaner's peril: the QMAP series is the most recent standard for deciding which local names should be actively dropped in favour of a single regional name, yet if the local formation has a stroppy author, and perhaps a band of supporters, the cleaners may find themselves in a fight. A low-level feud exists between those geologists who isolate and defend local strata, and those who seek to correlate them under a single name. Like any turf war, it has its own argot: the Splitters, at war with the Lumpers.

It's a somewhat arcane area of geology, but interesting because of the arrival of a huge new player — Zealandia.

The great sunken continent that lies espaliered either side of New Zealand now has a growing band of acolytes who will point out its hills and valleys, its ranges and basins, its thoroughly terrestrial greywackes and granites, its huge size. Such a rock mass, the seventh largest continent on earth, deserves a stratigraphic lexicon that might take not just New Zealand but the wider continent in its stride.

That was why I went along to a 15-minute presentation by Nick Mortimer in the last hour of the last day of the 2013 Geosciences Conference in Christchurch. It was titled *Litho2014: a Cambrian to Holocene high-level stratigraphic framework for Zealandia*.

Nick is a petrologist at GNS Science, Dunedin. I knew he once managed PetLab, the New Zealand rock catalogue that houses 193,000 rock and mineral samples, and that he was an adviser to StratLex.

He began his talk that day with a standard definition of stratigraphy:

'The description of rock bodies of the earth's crust, their organisation into distinctive, useful, mappable units. Used to establish distribution and relationship in space and time, and to interpret geological history.' He embellished that with a favoured quote from the American petrologist, Paul Krynine: 'Stratigraphy: The complete triumph of terminology over facts and common sense.'

No question, then, as to where his loyalties lay between the Splitters and the Lumpers, but as I listened on, I thought Lumper was not the right word for either him or the framework he was proposing.

If the Augean stables of New Zealand geology were knee-deep in local names and needed cleansing, then he might leave that to the Lumpers down in the bullock yard, for the changes he proposed were higher level than that. Enter, through the elegant front door of these stables, the Austral Superprovince, with subcategories to define and explain the Gondwana phase of New Zealand's stratigraphy. Enter also, at exactly the same level, the Zealandia Megasequence, with subcategories to define and explain the stratigraphy that developed during and after the separation from Gondwana.

The new high-level names, and the high-level subcategories of those names, were justified in each case, said Nick, as geologically concise and regionally useful for New Zealand, but would extend too across Zealandia, where more and more offshore continuations of onshore strata were identified and knitted into the economic life of the country. The new categories would act as a precise reference and search tool and were based upon easily identifiable physical phases of stratigraphic formation. He put up a cartoon to illustrate those phases, which were listed as follows.

The Gondwana phase of —

Growth *by accretion, or volcanic intrusion*

The Zealandia phase of —

Stretching *of Gondwana's hinterland as Zealandia readied for departure*

Break-up and Subsidence *as Zealandia broke free, then began to subside*

Immersion *as Zealandia all but disappeared under the ocean*

Collision *as a new fissure propagated through the sunken continent, and as subduction and strike-slip faulting began to uplift part of Zealandia — the drying New Zealand — from its submarine lodgement.*

I looked at the cartoon. Beyond a nod to the new Austral Superprovince and the Zealandia Megasequence, I didn't try to retain the names of any subcategories. They were good enough, but paradoxically they were good also because you could forget them and concentrate on the powerful physical sequences that separated them one from another. *Growth — Stretching — Break-up — Subsidence — Immersion — Collision*. Those had a dynamic logic. They were memorable. You could imagine the sequences storyboarded for a movie.

But one of the naming changes stuck. One change rang in the ears, and it was the tiniest change of all. Within the *Growth* phase — under the aegis, that is, of the Austral Superprovince — Nick noted that some scientists and publications still did not capitalise the geological entity 'terrane'.

— In this new order, he said, whenever we refer to a specific terrane, we capitalise the 'T', affirming its formal status.

The despised terranes of 1980. It had taken a while, but shepherded forward by Nick Mortimer at 2.35 pm on 27 November 2013, terranes marched into the Stratigraphic Lexicon of New Zealand with their heads held high, and their hats on, and why not, for they had a great story to tell.

The Californian office of the United States Geological Survey was based at Menlo Park just south of San Francisco, and in the early 70s three of Menlo Park's best geologists began to explore the subtler implications of plate tectonics. The three either already had links, or because of the work they did would soon have links, with New Zealand. They were, in effect, the inheritors of California's earlier contact with Trevor Hatherton.

They were field geologists. Davy Jones was the expert on the Franciscan Formation — the coastal mountain that ran either side of San Francisco. He was joined by the ophiolite expert Clark

Blake, and then by a third and younger geologist, still learning, David Howell.

Plate tectonics supposed the incoming plate pushed up coastal ranges at the continental margin, then returned the sediments shed by those same mountains. That purported to explain the predominantly greywacke Franciscan Formation, but Jones for one, and anyone who knew the range, knew also there was a lot more than that to the Franciscan. It had a lot of indistinguishable glop. It had melanges. It had cherts. It had ophiolites. Why?

Through to the early 1970s, plate tectonics had been driven mainly by geophysicists. They used aircraft. They used ships. They used instruments that yielded magnetic and gravity data. Their work was relevant to any model or map with a scale of 1:250,000 and up, and they'd done the work of mapping the spread of ocean floors, of measuring heat at the mid-ocean ridges, of sounding the depths of the trenches and the size of the fold belts. They worked in labs often, and wore white coats. The field geologists sometimes called them 'non-humans'.

The field geologists knew their territory at a scale of 1:50,000 and under. They knew it down to a scale of 1:1, standing scratched and bleeding next to the remote outcrop, wondering how come the wide belt of high-temperature low-pressure gneiss sat beside a high-pressure low-temperature belt of schist. The big idea was waiting to arrive, and it arrived first in the minds of the geologists who'd spent years in the field and knew the problems. It was a mind shift, a 360° spin, an overturning of one of geology's most basic questions — not why certain rock belts fitted together, but why they didn't fit.

Maybe plate tectonics didn't just pile up an accretionary prism. Maybe the strange rock bedfellows way inland of the coast were exotic visitors from afar. Maybe, way back in the Paleozoic or the Mesozoic, not just offshore sediments but a whole volcanic island came riding in, or a volcanic arc, or a fragment of some former continent.

The Menlo Park team looked for sutures. They looked for distinct crush zones, or even a melt, where large and alien masses might have docked against the North American craton. Docking was signalled often by the ophiolites. It was signalled too in the fault lines that separated California's distinct stratigraphies. The Menlo Park team named their fault-bounded units 'terranes'. Those parts of California without boundary faults, but with strikingly disjunct stratigraphies, they named 'suspect terranes', and they began to shape a new picture of Californian geology.

Terranes theory and its praxis got under way in California. Menlo Park was beginning to suppose a Paleozoic coast that had once ended in blue sea lapping against Nevada. Menlo Park was beginning to believe California arrived later, in fragments large and small — a wrack line of maybe six different terranes, and sometimes the great jumble of rock that came to be called a melange, chaotic assemblies that seemed inexplicable in a static world, but whose origins in a kinetic world were obvious every time you swept the kitchen floor.

Through the 1970s, the Californian terranes theory was firming up. The whole state, in effect, had been assembled from somewhere else. New Zealand picked up the excitement of terranes, and responded. Otago University's Chuck Landis had already begun to name New Zealand terranes in the late 1960s, naming with a small 't' simply to identify known geological rock units but without, as yet, California's specific awareness that the units could be far-travelled. Davy Jones and Clark Blake came to New Zealand in the early 70s, and that helped the terranes idea along, but it was still mainly an idea, and not a revolution.

It's easy to pick, in this decade, a reluctance to let go of the New Zealand geosyncline. It explained so much, and New Zealand geologists had believed in it for so long. Geosynclines could explain every rock type in New Zealand very simply. The hardest rock — gneiss perhaps — was held to be buried deepest within the syncline, maybe something over 30 kilometres deep, greywacke was something under 25 kilometres, and so on. And then they were raised up to the light. By 'earth forces'. By isostasy. By whatever. Why complicate matters by having those rocks sail in from some godforsaken part of the world when you could produce the rocks — every type — right here in your own backyard?

Even the diagrams of that era that specifically drew a downgoing plate under New Zealand sometimes drew also, high above the plate, a scoop in the continental shelf marked 'geosyncline'. And if plate tectonics itself was having that kind of trouble, any talk of terranes being the next step beyond plate tectonics was simply a step too far.

Besides, in science you couldn't just jawbone terranes into existence, you had to prove them, and that was a massive undertaking. There was good science being done, though, by the Geological Survey's Otago regional geologist, Graham Bishop, along with Landis. By John Bradshaw from the University of Canterbury, who was mapping a melange belt within the Torlesse rocks of his province, and that belt had definite tectonic origins. Doug Coombs and others of Otago University in 1976, and Auckland University's Bernhard Spörli in 1978, published papers that were significant identifications of New Zealand terranes. Both papers are detailed and persuasive. They name 'litho tectonic' or 'tectonic-stratigraphic' terranes, terms that signify delivery of distinctive rock units by an outside force, and the papers are clear what that force is — plate movement. The papers are brilliant in their way, insightful, yet cautious also, working from the field observations up towards the theory. But they can't resist labelling in their diagrams the 'geosynclinal province' (Coombs) and referring to tectonic process 'in the New Zealand geosyncline' (Spörli). Reading both papers, you're aware their authors are on the brink of a full terranes theory, yet it's also true, as the 1970s progresses, that the wider context languishes and the ghost of the New Zealand geosyncline still lingers.

In 1979, David Howell arrived in Dunedin. His USGS mission was to study the Torlesse rocks, New Zealand's great belt of greywacke, but he travelled across the North and South islands in the company of whatever New Zealand geologists would come with him, hitching rides with the barefoot chopper pilots of Fiordland, or burying the axles of 4WDs on muddy tracks, and he had a wider agenda. He was looking for the big faults, or the disjunct stratigraphy, that might advance the theory of terranes, and when he wasn't doing that, he lectured at Otago University.

Howell told his students at Otago that geologists were rare romantics who lived in a model of fantasy constrained by only the most meagre amounts of data. Take an idea, he said, and try to find the evidence that suits, for without an idea you didn't know what to look for. He likened a geologist to the Tibetan monk whose vision of the bardo — the stages of the soul after death — was so vivid, so eloquent, that those who listened, believed. Howell was vivid. He was eloquent. He talked a blue streak on terranes at the university, but the geological establishment beyond the campus remained unconvinced, both cautious with the idea and opposed to his scientific method. You took your big idea into the outdoors and gathered evidence to suit? You didn't. Scientific method in New Zealand demanded you gathered your evidence and analysed it before venturing your theory. Howell felt at odds with the departmental chair at Otago University. He felt at odds with Wellington. He offered a lecture to the Geological Survey's Wellington headquarters, but was turned down, and it took an invitation by the recent past president of the New Zealand Geological Society, Ian Speden, to get him to the capital. Howell talked, but the audience was sparse. A competing geological meeting had been called that day, and he wondered later if his audience had been deliberately decimated.

In 1980 Howell published his paper 'Mesozoic accretion of exotic terranes along the New Zealand segment of Gondwanaland'. He walked New Zealand back to the Mesozoic, over 200 million years ago, to Gondwana, and immediately he was addressing a minority audience. The NZGS was staffed almost entirely by Cenozoic specialists, and their expertise was New Zealand's soft cover rocks. Cenozoic rocks held a wealth of information for the sedimentologists and the macro- and micro-paleontologists, but what use was the Mesozoic? So very distant in time. So lacking in fossils, or earthquake return times, or pollens, or ice age evidences, or eruptive patterns, or anything, in short, that seemed relevant to the present day.

And then came a second black mark. The paper had a simple line drawing of Gondwana where ur-New Zealand featured as a dotted outline on the ancient coast, and that was acceptable, but offshore of Gondwana, in the ur-Pacific, was something you should never see in an academic paper. The ocean was truffled with a cavorting sea monster. Howell had sketched in a sea dragon with large scales and hooped nostrils. It was blowing smoke.

The third black mark was against his portrayal of the terranes themselves as far-travelled accretions or amalgamations. The proofs being offered were driven by that top-down wishful thinking more than from hard field evidence from below. And what about the New Zealand geosyncline?

Gone, and not even a fond farewell. In its place the idea of a convergent margin, an ocean plate that might bring in both subaerial fragments and submarine sediments at an angle, and which meant any terrane lodged within the New Zealand mainland might be thousands of kilometres distant from its source. Howell's paper implicitly accepted a spherical geometry of plates moving on a globe, and that geometry practically precluded any plate ever coming in at a right angle. The paper had a simple suggestive power. Ur-New Zealand had lain along Gondwana's south-eastern coast. The Gondwana craton was very old, 600 million years and more. The ocean floor that subducted constantly at its foot was young — nowhere older than 180 million years. That must have led to high-pressure metamorphic process along the south-eastern edge and to what Howell called 'the accretion of exotic terranes'. He defined terranes, and from his New Zealand field trips he listed four. One was part of original Gondwana. The others had come in from elsewhere.

The paper was a revolution — not of data, which was already to hand in many New Zealand geology papers, nor just of names, for the Torlesse terrane, for one, was already named. It was a revolution against any kind of fixity. Even the prevailing interpretation of the Torlesse as erosion off the Gondwana coast plastered directly back onto the coast by an incoming plate implied too much fixity for Howell. Coombs had already hinted that the sediments making up the Torlesse were unlikely to be local, and Howell affirmed it more clearly yet. The turbidity currents of undersea rivers might dump Gondwana sediments way out on the abyssal plains, and the delta fans from onshore rivers might strew them far out to sea. The incoming ocean floor would return them to Gondwana in due course, and in due course also, they'd become part of an accretionary prism building against the Gondwana coast, but if the incoming ocean floor plate was coming in at an angle to Gondwana, it would return the sediments many hundreds, or many thousands, of kilometres south of the original source. Or thousands of kilometres north. No one knew the paleo-angles. Thus David Howell: a small group of New Zealand geologists had done some powerful work on terranes during the 1970s, but no New Zealand author had ever written anything with the focus, brashness and assertiveness of his one short paper.

To gauge the effect of Howell's paper in New Zealand was to gauge also the frustration of the group of young, mostly South Island geologists. Call them the Young Turks. They were restless even before the publication in 1978 of the Geological Survey's great masterwork, the *Geology of New Zealand*, but when that publication did arrive, they'd looked upon the two-volume edifice and despaired.

There, mapped offshore from the North Island, was 'the New Zealand syncline'. There were the geosynclines again, prominent in the table of contents. The two volumes barely mentioned plate tectonics, or Gondwana, and certainly not terranes. You couldn't gainsay the *GoNZ*'s immense compilation of knowledge, but it had stalled at the Government Printer for years longer than it should. The main editor and head of the Geological Survey, Pat Suggate, had reputedly carted the manuscript onto the *Wahine* when the ship departed Lyttelton for Wellington on 9 April 1968, and as the story went, had come ashore amidst the bedraggled survivors on Seatoun Beach next morning clutching it to his chest. The *GoNZ* might have passed muster in 1968. It didn't pass muster in 1978. Better by far, muttered the South Island's Young Turks one to another, if the *GoNZ* had gone down with the ship.

Such was the story told to me, and sworn as true, by four separate people during my year of geology. Pat Suggate did come ashore at Seatoun that horrible morning hugging a geology manuscript, but it wasn't the *GoNZ*. I've left the story intact here, though, because it has wide currency, because it illuminates the discontent so memorably, and because it illuminates the half-truths and humour that keeps any urban, or maybe that should be earthen, myth alive.

Then came Howell's paper, and the astonishing data beginning to flow through from Menlo Park's geophysicists. Paleomagnetic study of Wrangellia, one of the largest North American terranes, suggested it had travelled from the tropics to 62° north. That was the equivalent of Fiji sailing past Invercargill on its way to Macquarie Island.

The Young Turks were not just pleased by Howell's work, they were galvanised by it. Their bond was the South Island high country, and they'd met in alpine huts at Desolation Spur, and Blue Ridge and the Garden of Eden. All of them were already awarded, or would later be awarded, the McKay Hammer, or would hold, as was John Bradshaw's honour that same year, the coveted role of Hochstetter Lecturer. Bradshaw taught at Canterbury University, Chuck Landis at Otago University, and the others were all Geological Survey staffers — Graham Bishop, Ian Turnbull, Peter Andrews and Guyon Warren. They were New Zealand bedrock specialists, Mesozoic specialists within the largely Cenozoic world of the Geological Survey.

Word went around. Howell had done well, and they could do better. They met within the old Gothic pile of Otago University and unrolled newsprint down a long table. They took up felt pens. They outlined a map of New Zealand. They drew in the Alpine Fault, and the splay faults that go out through the Hope Valley and Marlborough. They stood back, then drew in the faults that bounded the Dun Mountain Ophiolite Belt, the Murihiku Basin of Southland and the Torlesse. Easy.

In the 1970s, the dominant model for New Zealand's overall basement geology was of two parallel but easily distinguishable belts of metamorphic rock, the high-temperature and low-pressure rocks of the so-called western province, and the lower-temperature but higher-pressure rocks of the so-called eastern province. The boundary between the two belts, described as the median tectonic line, was a collection of rocks, different in kind from what were — simply expressed — the old metamorphic rocks of the west, or the young sedimentary rocks of the east. The rocks of the median tectonic line were igneous. This basic division of New Zealand's basement rocks existed then, and exists now, quite independent of terranes, but the problem was how those two separate provinces might be further divided into terranes. The Young Turks moved around the table. The easy bit was over, and now they began to argue the rest. Guided by the known large faults, they marked out the rocks of the older western metamorphic province into two terranes, one of them an original piece of Gondwana, the other, they believed, an ancient accreted terrane. Then came the younger eastern province that lay Pacificward of the median tectonic line. In a 14-hour marathon, fuelled towards its end by pizza and beer, the young geologists finally laid down seven more terranes in that younger belt, bringing the total to nine. The boundaries of those terranes have hardly changed since.

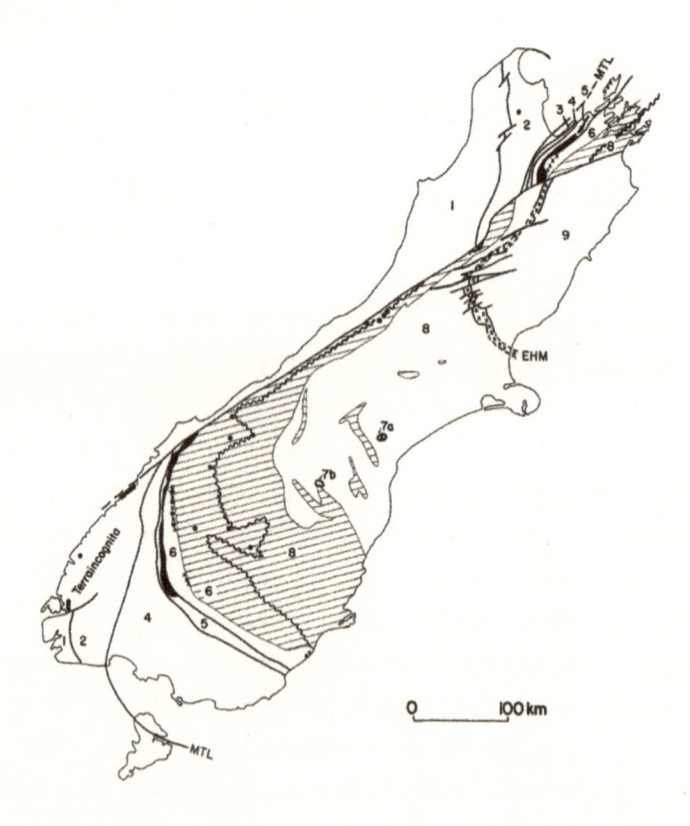

California became the centre of terranes study worldwide, and in 1983, Blake, Jones and Howell convened the Circum-Pacific Terranes Conference at Stanford University. Across from New Zealand came Bradshaw, Landis and Bishop, bringing with them their spruced-up terranes map that now had a striped belt added, not a terrain but a zone where the separate rocks of the Torlesse and Caples terranes had undergone deep burial and metamorphosis into separate sorts of schist. And down from the Klamath Mountains north of San Francisco came Nick Mortimer. The young Englishman from London's Imperial College was beginning the last year of his PhD. He could do the lab work to analyse rock type and ages, but he enjoyed fieldwork. He'd spent weeks in the Jurassic ophiolites and Permian volcanic arcs of the Klamaths, and he needed no convincing on terranes. He arrived at the conference and noticed the New Zealanders. They were all tanned and gaunt, and wore long shorts, and white socks pulled up just below the knee.

Graham Bishop picked up the new recruit from Momona Airport in a Land Rover and deliberately steered clear of State Highway 1, driving him back into town across the Taieri Plains, then Three Mile Hill Road.

The year was 1986, and the new recruit would recall wondering — *Yes, but where is Dunedin?* They topped out at 350 metres elevation above the city, and Bishop's surprise was sprung. In a sudden sweep, there was Dunedin. The new recruit would recall clearly that first bright expanse of blue ocean, the peninsula, the coloured city at its foot. They came down past the dry-stone walls and the fish and chip shop, and that side of it pleased the remnant Englishman in the new recruit, but the city itself lay in the clasp of an old Miocene volcano. The same dramatic Pacific Rim geology he'd known in California and British Columbia was here too.

The 27-year-old Nick Mortimer thought he might like Dunedin. Later he put it more bluntly. He loved the place, a home away from home. His base here would be the newly opened DSIR Geological Survey building in Cumberland Street, his colleagues would be a small tight team of mainly Mesozoic geologists, the library and café of Otago University were right next door, and his first fieldwork was scheduled to begin in what's commonly called New Zealand's most desirable destination. Nick had taken up a three-year scholarship to map the boundary between the recently-badged Caples Terrane and the Torlesse Terrane. The most prominent boundary between the two would turn out to be halfway up the Remarkables Range. The craggy jointed crenellations of the Remarkables are hard volcanic-arc-derived Caples schist, the wide skirts that slope away below are the softer and more slippery Torlesse mica schist. Nick Mortimer's job began right there, in Queenstown, where the flash architecture of the timeshares and the hotels was aligned across the reflective waters of Lake Wakatipu to exactly that mountain view. He stayed a few weeks, then moved out to four-wheel-drive and walk the long low tops of the Otago ranges, mapping the meandering Caples boundary over 300 kilometres through Otago province.

Other proposed terranes, meantime, came under further field scrutiny from the Mesozoic specialists in Dunedin and Christchurch. The proposed Brook Street Terrane was one of New Zealand's most astonishing pieces of geology, noted by Hochstetter as diabase eruptive rocks in the Brook Valley behind Nelson, and named after it, but was at its most impressive when it emerged, after offset by the Alpine Fault, in Southland. The Takitimu Mountains there were a part of it, so too the rocks around Riverton.

Chuck Landis in company with two American geologists knew Brook Street Terrane was a volcanic arc, a Tonga equivalent or a Fiji perhaps. Their paleomagnetic surveys suggested the arc had sailed in from latitude 27° south to dock against ur-New Zealand's

Gondwana Coast, then lying at around 62° south. That was a 3500-kilometre voyage across the ocean. With no human eyes to see it, still the new terrane seemed a hallucinatory traveller as it hove into view in its million-year approach to the coast, the black nibs and spears of it etched against a Permian sky. The study was questioned later for some of its procedures, but no better study has been done.

The sedimentary Murihiku basin lay immediately north of Brook Street Terrane. The Murihiku sediment was volcanic, and the Brook Street Terrane was volcanic. Surely the basin had simply filled with sediment from the eroding Brook Street, but John Bradshaw and others found the Murihiku basin had been fed by fresher, younger Triassic and Jurassic volcanic arcs. The basin held a million cubic kilometres of sediment, eroded off some great volcanic edifice. Reconstitute that million cubic kilometres into an original form, and you'd be looking at over 100 Taranaki-sized volcanoes. Where was the ruin? The geology sleuths who sought the provenance of the Murihiku Terrane would come to be called the Raiders of the Lost Arc, and they never did crack the secret at its heart, but the surfaces of the terrane continue to host a grateful and bleating present: the rolling Southland Downs, populous with sheep and lambs, cross-cut with flax-fringed tarsealed roads where stock trucks shift herds of dairy cows on Gypsy Day, and with fields full of winter swedes, sweetened by frost and fattened half out of the ground.

After Murihiku, the ophiolite belt had come in, and was established as the probable mid-ocean ridge of some previous plate impelled towards Gondwana and then firmly wedged into the Gondwana crust by the advance of the newly active Pacific Plate.

And so it went on, terrane collisions that were sometimes tens of millions of years apart, each different from the one before, until the last and greatest of them began to build. The Torlesse Terrane was vast. It extended for thousands of kilometres along the Gondwana coast, an accretionary prism where Gondwana's own offshore granitic sediments gathered and folded back onto the continental edge. Howell was right that the sedimentary source would prove elusive, though zircon analysis 20 years later showed ur-Queensland as a possibility. But whatever its provenance, the Torlesse was a thing of wonder that the earth could ever stand still long enough to build such a pile, and even with the occasional discovery of a swallowed seamount, or a limestone chunk, or a layer of pink chert, it remained mesmerising in its sheer monotony. But if you wanted a solid foundation, it was terrific. Here was the signature stone of the South Island, and the spine of New Zealand itself — the greywacke.

Soon after completing mapping of the Caples boundary, Nick Mortimer joined the Geological Survey's Petrology Section. Petrologists deal with rocks at a microscopic level. They will progress in the analyses they undertake from a hand lens, to microscopy, to mineral and geochemical analysis, and may finish by counting the decayed atoms within a zircon. In the course of which increasingly minute work, or so a lay observer might suppose, any petrologist might shrink sufficiently to fit into a pencil case, slide its lid back over himself, and say goodnight.

Nick Mortimer out running on the Otago landscapes. Skirting the matagouri. Assessing the right angle of descent through the gullies. Glancing at the small aids that keep you on course for a smart run: a thumb compass, a map with contours down to five metres, and the individual tors marked out. Consulting the detail, to navigate swiftly through the wide-open spaces of Otago's Waipounamu surface. An orienteer, running to a placing within the Otago championships.

Nick Mortimer would argue back that petrologists are like pathologists. They do analytical autopsies. They assemble forensic evidence, and they establish what happened. The big in the small. A petrologist who attends to the detail within a single thin section of rock can give you the provenance of a mountain range, or perhaps the structure of a continent. As it turned out, he was on the brink of doing just that.

In the mid-1990s, in company with his Dunedin colleague Andy Tulloch, Nick began to seek the provenance of New Zealand's huge and puzzling piles of old igneous rocks. Those rocks had been ignored in the first terrane map produced by Landis, Bradshaw and Bishop. Yet the Permian, Triassic, Jurassic and Cretaceous hard rock had shaped some of New Zealand's most memorable landscapes, from Fiordland's blank empire to the bony summits along Westland's Paparoa Range, and Southland's Longwood Range and Stewart Island.

The igneous rocks had been acknowledged in the Old Geology as the median tectonic line. As the New Geology took hold, they'd been marked up by Bradshaw as the Median Tectonic Zone, a holding term for rock masses that seemed likely, with a bit more investigation, to be proven as terranes. Radiometric dating work showed separate igneous masses with a wide range of ages, and the fieldwork showed a physical separation, often, between the plutons, and some faulting against the side of them. All those evidences favoured the theory that here too was a bunch of 1980s-style exotic terranes.

Nick Mortimer was a terranes man. He was amongst the first to get the concept, and he was amongst the first to get over it. He called the igneous masses by a tongue-in-cheek name that shook free of any previous association — the Median Tectonic 'Phenomenon'. It could be New Zealand's biggest and best terrane, but it didn't have to be. Like Howell, he was working a hunch, and he went after the evidence.

Nick turned the Nissan 4WD off the highway at Otautau and into the Longwood Range, picking his way through a maze of forest roads until the bonnet began to tilt towards Bald Hill (804 m), the highest Longwoods peak. We were going, in his words, to put our noses on the Median Batholith.

The Longwood Range is long and low, with substantial forests along its flanks dwindling to subalpine shrubs and then tussock and sometimes bogs above the treeline. The weather arrives often from the unprotected south coast of the South Island in full blast upon the Longwoods, and you don't readily enter the range without a strong motive. Back in the 1880s that motive was alluvial gold, discovered at Round Hill, as the Longwoods slope down towards the sea at Colac Bay. The Chinese built their own township at Round Hill, complete with a joss house, a hotel and food store. Chinese miners worked the auriferous grounds with cradles and matting for over 20 years before European joint stock ventures accelerated the mining in the early 1900s, employing the Chinese to build long water races across the Longwoods flanks, and to dig out holding dams for the water above Round Hill. With that powerful head, they directed their hydraulic artillery against the alluvial faces, washing them down to bare cliffs above and a slow-moving sludge below.

Back in the mid-1990s, Nick Mortimer had a strong motive. The Dunedin office had the power to write its own agenda, and he decided it was time to figure out what the igneous so-called 'Phenomenon' was. Andy Tulloch went to Stewart Island. Nick went into the Longwoods. They were hunting New Zealand's biggest terrane boundary, or something else, still big but unsuspected, overlooked. They were looking for the evidences that might test their ideas. Nick explored through the dense forests of the Longwoods so often and in such marginal weather he came to call it Mirkwood, but he got results and put them together with Tulloch's.

The two geologists found hornfelses, rocks whose original texture has been destroyed by heat from molten granites and gabbros. The hornfelsed rock belonged to Brook Street, the arc that had docked against Gondwana in the early Permian. Something very hot had come up inboard of that terrane. Howell's terrane faults had been marked out typically with crushed rocks lubricated with serpentine and clay, but the two geologists found no evidence of faults. They did find evidence of a heat sufficiently pervasive to weld the eastern and western provinces together. The ancient and smoothed domes and summits of that weld were not an accretion from the side at all. They were a prolonged and multiple igneous intrusion from below, a giant volcanic chain.

Nick turned a corner, and stopped the ute. He pulled on the handbrake.

— There's a feature on the corner worth showing you, Geoff.

We walked back in a light cold rain. A monolithic granite wall rose above us, a pluton from the Median Batholith sliced years back by whatever bulldozer carved the road, and inlaid in the granite was an orange oblong the size of a roof tile.

— It's a piece of Brook Street. A xenolith. A nice little set piece. We know from our dating that the granite is younger than Brook

Street. This chip has been engulfed by the granite, entrained and taken in, so it had to exist before the granite came along. And you can see the black alteration rind around the edge of it. The granite was still hot, and it's altered the minerals.

He put an affectionate hand on the xenolith.

— Brook Street came in first. When you see something like this in front of your face, you know the granite had to inject into it. A massive hot intrusive contact around two hundred and fifty to two hundred and sixty million years ago. It's not a fault contact.

We drove on up, unlocked the encased padlock on a heavy metal gate that blocked any rogue 4WDs from the summit, and stepped out into an icy wind and cloud.

On a good day the Bald Hill summit offers panoptic views over a distant Fiordland and the wide reaches of the Southland Plains. Seen from below, the summit with its tall telecommunications tower is a landmark for the farmers and townspeople of Otautau and Riverton. In the early 2000s, the Longwood tops became a brighter beacon yet, when shafts of lights beamed upwards in the night. A mineral exploration company moved in here and worked 24/7, test drilling for platinum.

We explored through the tussock and alpine shrubs, looking at the boulders.

— This is now gabbro, and layered gabbro can hold platinum, said Nick. It's much darker in colour than the granite, a lot more pyroxene and hornblende. All over the Longwoods it just seems to overlie everything.

He scraped away a bit of lichen, and traced a dark vein running through the speckled gabbro.

— These veins are basalt. We're underneath volcanoes here. Granite, gabbro, basalt. This was all molten rock in its day, but it cools at different temperatures so we're dropping out crystals, we've got mushes and things cutting each other. It's all very hot and mobile and it's all frozen solid for us to look at. The gabbro cools the slowest. It draws in water and things, so you get layering where the residual liquids are squeezed up and concentrate elements of interest. Platinum and other rare metals. That's what interested the mining company.

A Te Araroa direction sign stood nearby, and I went across to it. The trail climbed this summit north from Merivale Road and went on south across the Longwood tops towards Riverton, and the trail end at Bluff.

— You look cold, said Nick. Do you want to go back to the truck?

The icy flecks streaming towards us were steadily increasing.

— I'd say this is snow, it's good enough, said Nick.

The mission for the day was to hike over to Summit 794, three trackless kilometres distant. Before packing its bags, the mining company had done an aerial survey of the Longwoods, and their magnetic map showed another big gabbro outcrop around Summit 794. Nick wanted samples, but we sat in the ute and waited for the snow to either ease or worsen.

— Of course, you could tell your Te Araroa people, said Nick, that they're walking the back of New Zealand's greatest mountain range.

— Really?

— Give or take a few million years. Every mountain range on the planet is underlain by a thick crust. These granites and gabbros once had that sort of thickness. Andy Tulloch, my colleague, has coined the term Cordillera Zealandia for the mountain range that existed on top of it.

— Well okay. What height are we talking?

— There are ways of estimating that, by oxygen isotopes. It can be done but it hasn't been tackled yet as a specific head-on problem. But let's say, by analogy, the Cordillera Zealandia would compare with the Andes. Many of today's mountain ranges, the Himalayas, the European Alps, are pushed up by continent-to-continent plate collisions. The Andes are forced up above a subduction zone, and you'll see active volcanoes there as well as earlier compression, that sort of thing. It'd be much the same for the Cordillera Zealandia. The Andes, of course, is a very respectable mountain range.

Geologists live in a world that no one can see, but they see it. We were sitting in a truck at 46° south.

— There'd be perpetual snows? I asked.

— Oh, it's snowy, said Nick. We're south of seventy degrees latitude. We're inside the Antarctic Circle.

In 1999, Nick and the Dunedin team published 'Overview of the Median Batholith, New Zealand: A new interpretation of the geology of the Median Tectonic Zone and adjacent rocks'. It outlined a bold conceptual change that took most of the scattered igneous piles in the South Island and turned them into a single thing. A batholith is formally defined as any igneous rock body with an area of 100 square kilometres or more. The Median Batholith was 100 times that size. The paper proposed a magmatic arc that had been active for at least 250 million years. It varied in its radiometric dates, but only because it was so long-lived and its activity had come in pulses. It was faulted along its margin in the north, but only because slippage along the Alpine Fault — a Cenozoic movement and not a Mesozoic one — had caused secondary faults long after the batholith's original intrusion. The apparent separation of the igneous massif was by cover rock only and did not compromise the unity of the batholith's many contiguous plutons.

The Median Batholith was an Andean-style magmatic arc, and that implied a set of features any geologist would recognise — an offshore trench running parallel to the coastline, subduction pressure against the coastline and, at a distance inboard of the coast, a line of mountains raised by compression and volcanism. The volcanic tops worn off since. The batholith.

Geology repeats itself. The same thing — writ somewhat smaller

— was happening right now to New Zealand, the Pacific Plate descending off the eastern coast, and the magmatic arc rising in parallel, 120 kilometres north-west on an axis stretching north from Ohakune. Except there were two big differences. The Median Batholith magmatic arc had been extinct for 100 million years. And it was huge. The Cordillera Zealandia.

We sat watching the snow, and envisaged the great mountains. Bald Hill, lesser in height than its illustrious predecessors, was nonetheless rapidly drawing a noble cloak of snow about itself, and we agreed to abandon our tramp. Back on the highway, Nick drove west, and took the turn to Monkey Island. Once at the picnic ground there, he shouldered his pack and moved off. The tide was low, and I wandered across to the distinctive pile that was the island. Ngai Tahu had inducted the island into myth. They'd named it Te Puka o Takitimu, the anchor stone of the ancestral *Takitimu* canoe, wrecked long ago in Te Waewae Bay.

The later settlers around Te Waewae Bay sheltered their craft in the lee of the island and would haul the boats onto the hard by working the long lever of a monkey winch. It was easy to imagine how the island received its next moniker.

Ngai Tahu's imagined anchor suited the topology of a wide region. In myth, the Takitimu Mountains 60 kilometres north of here were the canoe, and the Hokonui Hills, 70 kilometres north-east, were the sinking craft's desperate bailer. The Maori name, half sunk in the landscape, served the dual purpose of remembrance and orientation. The Pakeha name was entirely local, a vivid little local vignette.

I climbed to the island's lookout and it was a thoughtful place. Two French women came up and we chatted while I watched Nick come slowly back along the shoreline, still casting about. I went down to meet him as he emptied his pack beside the truck.

Smooth grey stones fell out, and I asked him what exactly he was collecting.

— Rock with special thermo-chemical properties. What I've tried to do is pick the gabbros. The pyroxene gabbro. It's anhydrous, no water. Then there's a couple of other things to watch. It must not be fractured. No lines of weakness. All these cobbles are very tight. The second thing is there mustn't be any hydrous minerals. They tend to expand and contract a lot more with heat. These are nice, and hopefully they'll do the trick. I can't be held responsible for any explosions.

— And what was the trick then?

— A colleague of mine has built a sauna. He wants the right kind of stones.

We drove on to Colac Bay. The contact between the Median Batholith and the Brook Street Terrane lay midway along the bay, but the surf and the sand overrode it, and we carried on west to Tihaka Beach and began to hike Te Araroa's track through to Riverton. We walked across long-shore drifts of orange pebbles from the batholith granites further west, over a grassy headland, and as we descended to the coast again, I waited for the Brook Street Terrane to impress me, volcanic, and harsh with it. What I didn't expect was the silky blue rock that flowed out from beneath the sheep paddocks overhead in platforms dimpled with smooth potholes, platforms that stepped down in poetic shapes to the surf that swirled at their feet.

I walked along one of the platforms, and Nick called up from below.

— You're standing on a two hundred and seventy million-year-old sea floor. You can do the same with any sedimentary rock, of any age, but Permian is getting back there in geological time. It's fairly respectable.

We looked for the tattooed rock, the trace fossils that Maori call mokomoko, and it took some time but we did find them, wetting down the face of a rock layer to reveal 270 million-year-old traces of burrowing worms that took the purple of their starting layer down into the pale depths beneath, or dragged their pale layer into some purple darkness below, working their primitive palette, thousands of small finger painters out of the Permian.

— Very subtle. No great terrane or big-picture thing here, Geoff. Just details. But they break the monotony.

What I didn't see was any obvious volcanic forms, and Nick agreed. We were walking along the apron that spread out from an eroding volcanic centre, the sandstones, the siltstones and mudstones. At least half of the Brook Street Terrane was that kind of sedimentary rock, but its source was volcanic, its colours derived from basalt. On the way back to the truck, I picked up an interesting pebble, hard and glassy, and passed it along for analysis. Nick gazed at it through his hand lens.

— It's a very fine-grained sandstone. This was a sedimentary rock, but like many of them around here, it's now metamorphic. We're only a few kilometres from the Median Batholith, which was a big hot thing, and these are metamorphic. This is hornfels again. It's starting to take on a flinty texture because of the temperature. Microscopically it's starting to crystallise. If you had this in thin section, you'd see new metamorphic minerals growing out of the original minerals.

The metamorphic aureole. I looked back towards the looming bulk of the Longwoods.

— Really? How far can those things transmit heat?

— In round numbers, molten granite and gabbro sits at about a thousand degrees centigrade, and we have simple heat conduction equations. Typically you can expect to see thermal effects from one hundred to one thousand metres out from a decent sized pluton.

With the right mudstone, heat could also do in this part of the world what deep burial had done up at Nelson. I picked up another interesting pebble, and passed it across. Nick looked at it, licked it, and looked again through his hand lens.

— Argillite.

We ended the day at Kay Roughan's house in Riverton. Kay had an enthusiasm for geology that dated back to childhood adventures along the Milford Track and picnics out in the Otago landscapes. Her home in Riverton was often host to geology expeditions, and the pink lamingtons served at her morning or afternoon teas had become duly famous alongside the sausage rolls, the wide choice of dips and crackers, and the coffee. An Australian team had used her house as a base when it launched an investigation into the volcanic Solander Island just offshore from Riverton. The Australians had gifted her a Solander rock, and Kay had placed it on a special shelf, alongside other rocks with stories attached. The Solander rock sat alongside a flint nodule from a dredging expedition, and a piece of Bluff conglomerate that made clear her unequivocal stand on the Oligocene drowning. She'd named the conglomerate Kia mau ki a koutou mahunga ake ki te wai. Keep your head above water.

Next morning we took the short drive from Kay's house down to Howell's Point. Maori called it Taramea, and in Nick's opinion it was the best place in New Zealand to see the Brook Street Terrane. Two black oystercatchers with long red bills patrolled the shoreline with *peewee* cries, and the swell rolled down the side of Brook Street promontories, leaving them wet and gleaming.

The swell and the sand had done the work of fresh bedrock exposure, and as the wading birds advanced and retreated with the slow swell of the surf, we did the same, looking at the various sections of it. The sandstone shelves we'd seen at Tihaka were here too, but this was a different Brook Street entirely, the sandstone shelves surmounted now with ramparts and crenellations.

— When the tide's out, the kids play castles in them, said Kay.

— You can't quite make a volcano shape out of it, said Nick. The top has gone. Like the Median Batholith, it's had its head chopped off.

Two hundred and sixty million years of erosion had left us with only the basement plumbing of the volcano. We could see the stacked pillow lavas, see where the sheets of once-viscous lava had carried their rubble of broken pillows and cobbles slowly onward, and had then been, by the overwhelming ocean, stopped cold.

— The green colour, said Nick, petrologist to the fore, is when they stew in their own juices for a bit. That's when you grow the green chlorite, the green pumpellyite and green epidote minerals out of the basalt glass.

He took out the hand lens to look at the lavas.

— This one is very fine-grained. It's cooled quickly on the outside, and we'd say a chilled margin. And these ones have got gas bubbles that have been filled in with white and green minerals. They're round, and the dark green mineral in the middle is pumpellyite. Well, probably. You can't tell with a hand lens, it's a microscope job.

We stood back and widened the angles. We picked out by its dark grey colour and straight lines a horizontal chimney that cut through the chaos of green. A dike, injecting molten rock at speed, cracking through the hardened effusions around it.

— This would have been feeding lava flows higher up, said Nick. It's probably a little less altered. Probably a little bit younger. So it stands out. All of this is building towards the size, say, of an undersea volcano in the Kermadecs. Those are five to ten kilometres across. Quite large things, and we may be seeing the very edge of something similar here.

Some kind of selvage on a huge and knobbled tapestry that

went on and on. The Taramea rocks ran the eight kilometres around the coast to Tihaka and the Permian sea floor there. At Tihaka the volcanics had been mostly eroded away, but nor was that necessarily the end of it. Sink a shaft through the purple sea floors there, and there'd be more pillows and breccia below, and below that another sedimentary floor not yet exhumed, and below that another. The Takitimu Range, an uplifted block packed with these same volcanics, stood 1500 metres and more above sea level. Whatever it was and wherever it came from, the Brook Street Terrane wasn't small. From this shoreline it ran away east under Invercargill and parts of Bluff, and it ran back behind us west as far as the Alpine Fault, whose mighty tectonics had smeared it, and from which it emerged again in the Brook Valley behind Nelson, tracking away undersea and on under Taranaki before disappearing out across Zealandia.

The surf came strobing down the side of the shining promontories. The wind blew. The sun had vanished behind heavy cloud. A shag flew past, holding level to the horizon.

Kay faced inland and shaped a symbol with her hands.

— I want a big gateway.

— You mean a triumphal arch over Highway 99, said Nick.

— Kay, you want grandeur, I said.

— Yes. It deserves all the grandeur it can get.

We drove to Te Hikoi Southern Journey — the Riverton Museum. You could smell the tidal wrack of a seaside town. It was a Sunday in early May, the duck shooting season was just getting under way, and the duck shooters were in town with army-style camouflage, ammo belts and blackened faces, arguing the effective range of steel shot. It was illegal to discharge a shotgun within the town limits and the ducks knew that. They flew high until they reached the town limits, and then we could see them coming in to land close to the estuary bridge. The hunters in the town and the maimais out in the estuary bristled with indignation.

The museum was compiling a geology collection, and Nick was the main adviser. He was known here, and a short queue formed to take advantage of the visiting expertise.

— I found this stone on top of Putauhina Island, said the first rock hound. The muttonbird island. The soil's like potting mix up there, and you don't find stones, but right up on top of the island I found one.

— Right. A gastrolith, said Nick. A seal took it up there, probably.

— Yes, but it's heavy. It's almost like a meteorite. Want to have a look at it? I'll nip home and come back.

Word went around, and either they were themselves geologists or they simply enjoyed the magical incantations, for they held out their rocks and Nick didn't simplify his responses, and they retired, pleased. He took each rock, examined it with the hand lens, and identified them all one by one.

— This looks like good pristine hornblende on this surface here. You're close to the contact alright.

Or again.

— Chilled gabbro. On balance, without being totally sure, I'd say it was on the plutonic side of the contact rather than the metamorphic side.

Or again.

— This one, I'd call a diorite porphyry.

— Right, said the diorite's owner. I thought I'd carve it. It's got such a nice tight texture.

Carole Power, the museum manager, turned from Nick and greeted me.

— Last time I saw you was at the pub Dusty and I owned at Colac Bay. You turned up, muddy and dishevelled...

— And bleeding, as I remember it.

Route testing for Te Araroa in 2003, I'd come 12 kilometres along the flank of the Longwood Range on the service track beside Port's Race. The track had been overgrown and strewn with windfall, but underneath it held to a gentle descent, no more than four metres in every kilometre. A hundred years before, the Chinese had dug the adjoining Port's Race beautifully, tunnelling through many of the spurs, boxing it to cross the gullies, and supporting the crossing on wooden trestles. The racemen in their isolated service huts corked their letters into bottles and sent them floating down the race for later delivery. On calm nights, hut to hut, they floated off lighted candles through the dank forest. Gently delivered also by such small lighted rafts, or so the story went, was the opium.

Since that walk, the track had been cleaned up by the Department of Conservation, and Te Araroa Southland Trust had built a high-standard access track to join it, and to traverse parts of the old goldfield. Te Araroa was well known here, and I was scheduled to give a talk that night on the track.

I set up amidst the museum's display cases. Rank on rank of Maori adzes were on show, many of them shaped from the distinctive grey Southland argillite quarried from Tihaka and Colyers Island in Bluff Harbour. I had a digital projector, a laser pointer, and on-screen 3D maps that showed the trail winding through the landscape as a thin gold line. It was a talk I always ad-libbed but, equally, I'd given it many times before, and I found myself surprised at how it had begun to change.

— The trail begins here, at Cape Reinga. We start the journey on old ocean floor. Basalt. It's part of what's called the Northland Allochthon. It came in from elsewhere, right Nick?

I threw it to Nick. He was out there somewhere.

— That's right, Geoff, came the Voice of Geology out of the dark. The allochthon is basically a thrust event, a low-angle thrust at the base of it and you've got compression there, which is what a thrust fault is.

My year of geology. The laser's red spot onscreen had become less doggedly linear, more discursive, darting sideways to seek out the unseen trench that drove so many changes within the landscape. Through Northland, across the volcanoes of Auckland, the volcanic plateau, and out of the darkness, the Voice of Geology sometimes rose, amplifying the finer points. The Whanganui River, and Wellington, so to the South Island and the sunken block of the Sounds. The laser's red spot greased up and down the 3D landscape to show the line of the Alpine Fault and on through the Longwoods to Riverton, Oreti Beach, Invercargill to Bluff. I brought up the pictures of Te Araroa walkers in the joy of completion, swinging off the AA signpost there, of Te Araroa hikers' shoes dangling from the direction sign that pointed back to Cape Reinga, and I ended with a favourite quote from the writer Frédéric Gros — 'Evening will come, and the legs will have ended by engulfing, in small bites, the impossible distance.'

— Watch your eyes.
We'd been walking along the track that brings Te Araroa to its terminus at Stirling Point, when Nick left the track to scramble down to the base of a cliff. Fifteen metres above us, a rough mixture of cobbles was pasted onto the cliff, flock wallpaper upon one of the last ramparts of mainland New Zealand. Nick climbed towards it, and swung his heavy petrology hammer.

The chips flew, the sample broke away. Nick tested it with hydrochloric acid, and looked up.

— It's called the Foveaux Formation, part of the Oligocene shoreline. It's an unconformity, a geologically significant contact and boundary in its own right. It's literally clinging to the cliff. Another few hundred years of coastal erosion and it'll be totally gone. I've gone for a piece where there's lots of matrix. I haven't seen this lithology before. It's possible there could be something useful for extraction, like a microfossil, a radiolarian maybe, or a foram. Things that you can't really see.

An unconformity. The black gabbro of Bluff Hill was part of the Median Batholith, 245 million years old. The Foveaux Formation was somewhere between 25 and 30 million years old, part of the famous Oligocene drowning story. One overlaid the other, but if you touched the margin where Nick had sledged off the sample from its underlying rock, you put your finger on 200 million years of missing geology.

Not too many clues here then, though I'd met with enough geologists by now to know that, however truncated, they'd get something out of it. The rocks always talked. Dependent on what you wanted to know, the clues were everywhere.

The Mesozoic geologists, I thought, were writing a great detective story. Like most detective stories it was written backwards, starting with the dumb fact of a couple of beautiful South Pacific islands which just happened to have a continental geology, alps and volcanoes, at deeper levels a greywacke spine, a volcanic arc that came to visit and stayed on, a red slice of ocean floor shunned by trees. Amidst the discomforts and also the fun of their field trips, the Mesozoic sleuths had gathered the lines of evidence, wound the story back, put their theories to the test and argued amongst themselves. They were sometimes wrong but always curious, and they always, since the days of Alexander McKay and Harold Wellman, moved the story on. Nick probably saw it that way, for as we walked along the Foveaux track with Bluff at our backs, he said

— Even this single hill, Geoff. You're building on generations of geological work that enables you to localise it, and date it as Permian, and put it into the context of adjacent terranes.

Nick was known for his collegial generosity. A recognition of the thousand small bites that in science might progress towards some new insight. I recalled a conversation back in the Longwoods when we'd been driving along bumpy forest roads, in the rain, and, as can happen within the slingshot pleasures of travel, the words had seemed charged with magic.

'It's a big-picture scale,' he'd said. 'It's seeing the woods for the trees. Seeing the terranes and the batholiths. Nice big simple concepts that go across New Zealand, and you're not setting aside but summarising all the detail that by itself is almost indigestible. And it's huge.'

Because of Nick's expertise in petrology, he'd written some of the programmes for the offshore dredging of Zealandia. He'd been aboard those ships, and as the truck bounced along, on its Longwoods trajectory, he'd kept on with his recollection.

'When you're working offshore, because the samples you get are so few and far between, and because they're so expensive and there's been such effort to get you there, they take on a significance or relevance far beyond the size of the scruffy wet rock that you're holding in your hand.'

The scruffy wet rock. Geologists gained their evidences by whatever means. They wrote the story backwards into deep time, and I read it forwards. They helped light up the slow movements of the earth, and I'd had moments during the year when I'd seen the big mute forms of it — let's say — stamping on the spot. No actual movement — leave that to the wind and the trees — but whole horizons, dark and lumpy here and smooth and grey there, that seemed to quiver with ponderous momentum.

We walked on around Bluff Hill. Stewart Island stood on the horizon as a low shadow and we walked finally back into the everyday world of cars at the terminus of State Highway 1 and the yellow AA signpost that signalled the end of Te Araroa. The trail ended here.

— Yep, it's kind of nice to see the road sign, said Nick. The southern end of the country.

— But the geology doesn't stop, Nick, I said.

— You're right. The Median Batholith goes on from here,

Stewart Island is part of it, and then it goes on undersea and, we think, surfaces again at the Bounty Islands.

— Okay.

There went the Voice of Geology. It didn't stop with anything as obvious as land's end. It'd held in its hand one scruffy wet rock, and by that weight felt the geological enterprise expand through all the arcane skills and all the generations of its practitioners, their persistent tapping against the cryptic planet.

I'd spent a year amidst the vastness of it. To get anywhere near the occluded process of the planet at its work, you had to open your mind right up, transcend common sense, expand your reach towards the wonder and the madness of the gods.

Time to go. We shrank back to the size of men, got into the Nissan truck and began the drive back up State Highway 1 to Dunedin.

ACKNOWLEDGEMENTS

I was awarded a six-month Ursula Bethell Residency at the University of Canterbury in 2013 and Christina Stachurski, Paul Millar and Patrick Evans of the university's English department welcomed me in, and gave the book a great start. A generous Copyright Licensing New Zealand Writer's Award later that same year helped keep the book moving. Nicola Legat, then publishing director at Random House, and Dame Anne Salmond were supporters throughout, so too were Russell and Judi McKenzie. Tony Reid was a first reader and had valuable suggestions.

I acknowledge and salute the American writer John McPhee, whose five books on American and world geology, later gathered under the title *Annals of the Former World*, signalled to me as far back as the 1990s that a literary look at geology is possible, and were an ongoing influence. His books offered valuable conceptual categories. I adopted his 'Old Geology' and 'New Geology' as one instance, which is an obvious enough division with plate tectonics at its watershed, but which also happens to describe that transition succinctly, and hence seductively. I acknowledge also Elizabeth Kolbert, author of *The Sixth Extinction* and a relevant article on the French scientist Baron Georges Cuvier in the *New Yorker* of 16 December 2013; also M. J. Rudwick's books *Worlds Before Adam* and *Bursting the Limits of Time*. Both writers have stepped outside the usual Anglo-centric view of geology's pantheon to refurbish Cuvier's reputation. Rudwick is the most recent and now the preeminent scholar of the nineteenth-century geological period and was useful also as part of my research on Mantell and Lyell. My accounts of Alpine Fault history, and of Auckland's volcanism, use material that I published first in *New Zealand Geographic*. The last chapter uses material from Heather Halcrow Nicholson's excellent PhD thesis, *The New Zealand Greywackes*, though I conducted my own interviews, and my emphasis is different from hers. The suggestions of bulk and speed of the Maungataketake phreatomagmatic eruption at Ihumatao are taken from a paper, then unpublished, by Brittany Brand of Boise State University, Darren Gravley from Canterbury University, et al. Most of the other papers I've used are acknowledged within the text.

The four lines of poetry on page 64 are taken from Janet Frame's poem 'The Pocket Mirror', first published in 1967 by George Braziller (New York) and are used here with permission from the Janet Frame Literary Trust. The lines from James K Baxter and Les Murray on page 207 are from the poems, respectively, 'Nor did I ask' and 'Ernest Hemingway and the Latest Quake', used with permission. And the two lines from Rex Fairburn's 'To a Friend in the Wilderness' on page 40 are reproduced with the permission of Dinah Holman. Sir Robert Clerk holds copyright on John Clerk of Eldin's 'Unconformity at Jedburgh' and the drawing is reproduced with his kind permission. Thanks to Kim Ollivier for GIS calculations that helped give the Northland Allochthon some kind of comprehensible scale, and to Trevor Butler for conversations on engineering standards in earthquake country. Thanks also to Roger Smith of Geographx for his generous offer to supply the book with New Zealand landscape profiles, and to Liz and Pete Watkinson for use of their Leigh bach.

The many geologists I spoke to were always generous with their time, especially Garry Carr, Scott Nodder, Simon Nathan, Jarg Pettinga and Robert Crookbain, who gave early support, and Mike Isaac who, after we'd finished our Northland trip, helped hand me down the line to other geologists. He was then a first responder when the questions became, occasionally, arcane, and remained a good friend to this book.

The publishing director at Penguin Random House, Debra Millar, and non-fiction publisher, Jeremy Sherlock, encouraged me through a final draft of *Terrain*. Matt Turner was an excellent copy-editor, and Jeremy and project editor Anna Bowbyes then helped me through the many details of bringing the book to press.

As always, Polly, Irene and Amos stayed alongside, personally and by social media, to support and tease me along. Miriam read the text as it emerged, her observations acute and sometimes directly challenging but, as with everything, delivered with love.

IMAGE CREDITS

CHAPTER 1: NORTHLAND
page 9: Cape Reinga lighthouse, Amos Chapple
page 33: Northland Allochthon, Mike Isaac

CHAPTER 2: AUCKLAND
page 47: Motukorea looking westward to Auckland, Mark Meredith
page 48: 'Sketch of the Kings from the South', Felton Mathew, courtesy Auckland Libraries
page 50: Melted mantle rock beneath Auckland, courtesy Nick Horspool (adapted from Horspool, N. A.; Savage, M. K.; Bannister, S. C. 2006. Implications for intraplate volcanism and back-arc deformation in north-western New Zealand, from joint inversion of receiver functions and surface waves. *Geophysical Journal International*, 166 (3): 1466–1483)
page 55: Aerated basalt rubbing, Warwick Freeman

CHAPTER 3: EAST COAST
page 71: East Cape and Whangaokeno Island, Rob Suisted/Nature's Pic
page 75: USS *Eltanin* Voyage 19 Magnetic Profile, East Pacific Rise (adapted from the original)
page 107: Accretionary wedge on East Coast, GNS Science (seismic image adapted to show Pacific Plate interface)

CHAPTER 4: TAUPO
page 113: Taupo from Motutere Bay, Chris Warring

page 141: Distribution of ballistic impact craters from 6 August 2012 Upper Te Maari eruption, Rebecca Fitzgerald, University of Canterbury

CHAPTER 5: WELLINGTON
page 143: Te Papa — viewing the Iguanodon tooth, Woolf Photography

page 148: Downtilted terraces at Baring Head, Charles Cotton, taken from *Geomorphology: An Introduction to the Study of Landforms,* 7th edition, 1958

page 158: Unconformity at Jedburgh, Borders, John Clerk, courtesy Sir Robert Clerk

page 162: Iguanodon tooth (life-size), courtesy Museum of New Zealand Te Papa Tongarewa

page 169: Microtopographic map, dextral movement on Wairarapa Fault, courtesy Tim Little and David Rodgers

CHAPTER 6: MARLBOROUGH SOUNDS AND THE RED HILLS
page 171: Red Hills Ridge, Geoff Chapple

page 176: Mountain Mule logo, Miriam Beatson

CHAPTER 7: WESTLAND
page 197: Robinson Creek, Westland, Geoff Chapple

page 219: Pseudotachylite, Geoff Chapple

page 224: Alpine Fault movement: 20 million years BP and the present day, GNS Science (adapted from the original)

CHAPTER 8: SOUTHLAND
page 227: Southland Downs and Takitimu Mountains, Kea Photography

page 240: Early terranes map, John Bradshaw, Chuck Landis and Graham Bishop (adapted from the original)

page 245: Milford Sound looking north-west from Freshwater Basin, John Buchanan, courtesy Hocken Library

ABOUT THE AUTHOR

Geoff Chapple founded Te Araroa, the New Zealand-long walking track that opened in 2011. He is an author, journalist, playwright, occasional musician, and was once the librettist for an opera. He has won prizes for his movie scriptwriting and his journalism. Chapple's six books include a biography of Rewi Alley, the New Zealander who founded peasant schools in China and took part in China's 1949 revolution. He also wrote, from a protest viewpoint, a people's account of the divisive 1981 Springbok tour of New Zealand. In 2003, he won the Environment category of the Montana New Zealand Book Awards with his book *Te Araroa: The New Zealand Trail.* His stage play *Hatch* was produced by the Auckland Theatre Company and toured New Zealand and Tasmania through 2007–10. Since 2012, when he stepped down from the leadership of Te Araroa, he has returned to journalism and writing.

ERA	PERIOD	EPOCH	Ma
CENOZOIC	Quaternary	Pleistocene	
	Neogene	Pliocene	5
			10
		Miocene	15
			20
	Paleogene		25
		Oligocene	30
			35
			40
		Eocene	45
			50
			55
		Paleocene	60
			65

ERA	PERIOD	Ma
MESOZOIC		80
	Cretaceous	100
		120
		140
	Jurassic	160
		180
		200
	Triassic	220
		240

ERA	PERIOD	Ma
PALEOZOIC		260
	Permian	280
		300
	Carboniferous	320
		340
		360
	Devonian	380
		400
	Silurian	420
		440
	Ordovician	460
		480
	Cambrian	500
		520
		540

LOCATION MAP

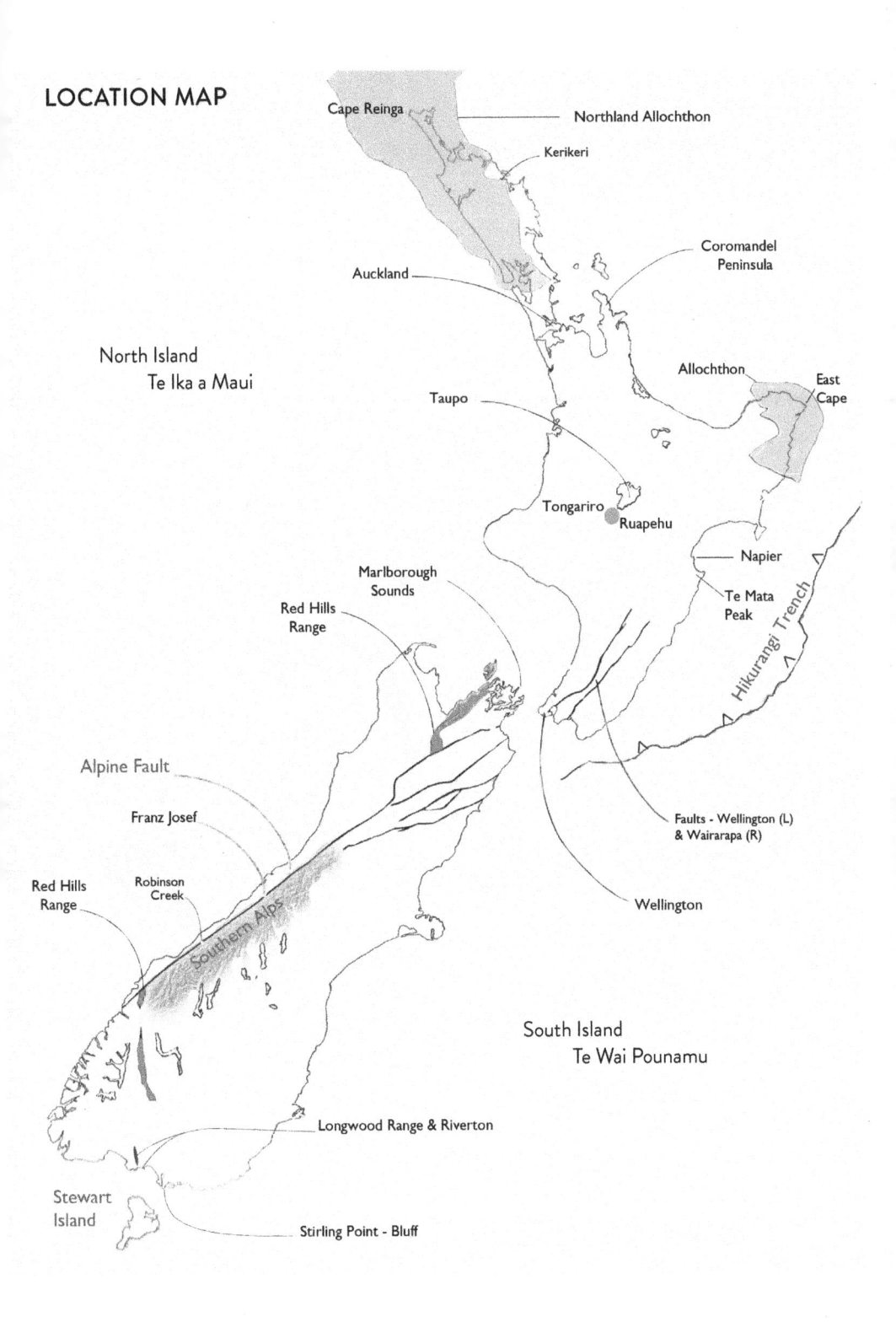

North Island
Te Ika a Maui

- Cape Reinga
- Northland Allochthon
- Kerikeri
- Auckland
- Coromandel Peninsula
- Taupo
- Allochthon
- East Cape
- Tongariro
- Ruapehu
- Napier
- Te Mata Peak
- Hikurangi Trench
- Marlborough Sounds
- Red Hills Range
- Alpine Fault
- Franz Josef
- Faults - Wellington (L) & Wairarapa (R)
- Red Hills Range
- Robinson Creek
- Southern Alps
- Wellington

South Island
Te Wai Pounamu

- Longwood Range & Riverton
- Stewart Island
- Stirling Point - Bluff